Mid othian Libraries

90 D0553426

This book is to be returned on or before
the last date stamped below.

7 MAY 1974   2 3 MAR 200

21 MAY 1974

-1 JUL 1974

23 JUL 1974

-5 SEP 19

ISLANDS BY THE SCORE

Part of the Gannetry on the weathered sandstone cliffs at the Noup of Noss, Shetland

ISLANDS BY THE SCORE

Alasdair Alpin MacGregor

**With thirty-five illustrations,
twenty-seven of which are reproduced
from photographs by the Author**

5/2832/441.12

MIDLOTHIAN
COUNTY LIBRARY

LONDON
MICHAEL JOSEPH

First published in Great Britain by
MICHAEL JOSEPH LTD
*52 Bedford Square
London, W.C.1
1971*

© 1971 *by Alasdair Alpin MacGregor*

*All Rights Reserved. No part of this pub-
lication may be reproduced, stored in a
retrieval system, or transmitted, in any form
or by any means, electronic, mechanical,
photo-copying, recording or otherwise, without the prior permission of the Copyright
owner*

7181 0860 4

*Set and printed in Great Britain by
Tonbridge Printers Ltd, Peach Hall Works, Tonbridge, Kent
in Plantin ten on twelve point on paper supplied by
P. F. Bingham Ltd, and bound by James Burn
at Esher, Surrey*

Preface

IN HIS preface to *The Enchanted Isles* Alasdair quoted the saying that 'every book is, in an intimate sense, a circular letter to the friends of him who writes it'. It is with diffidence that I add this preface to *his* letter to *his* own personal friends.

Several of the chapters in *Islands by the Score* represent the culmination of many years of frustrated attempts to reach islands he particularly wished to photograph and write about. When we first met in the early 1950s he would say at frequent intervals, 'I *must* get to the Treshnish Isles'. At that time I had never even heard of them. Not until 1966 did he manage to fit in a visit to Mull. We arrived there on Friday, October 7th, and the next morning rushed over to Fionaphort, where we learned from the ferryman, Mr. Alastair Gibson, that a party from Iona were planning to hire his converted drifter for a visit to the Treshnish. We booked in at the Argyll Hotel on Iona for the following day and returned to Craignure to collect a few belongings. At the time the weather was calm and sunny; but as we drove across Mull on the Sunday afternoon, the wind changed, and by that evening, as we sat by the wood fire in the Argyll's cosy sitting room, it was clear that the Treshnish Isles, for the time being, were unattainable. After a frustrating week of wind and rain Alasdair decided it was no good waiting any longer; we would leave the following day. But early the next morning we were awakened with a message that the wind had dropped and it was fine. By eleven o'clock we were cruising off Staffa, on our way at last to his longed-for Treshnish Isles. By the early afternoon the sea was like a mill-pond, the sun was hot, and there was good cloud; apart from a slight haze the day could not have been more perfect. We left Iona the next morning well satisfied.

Exactly a year ago today, Alasdair stepped ashore at the landing place at Kirk Haven, on the Isle of May, in the Firth of Forth. It was the realisation of yet another long-standing ambition; for although he had landed on the most remote and isolated islands off our coasts, he had not, owing to weather and circumstance, achieved a visit to the May. In July, 1969, when in Fife on our way back from Orkney, he had managed, after two abortive visits to Anstruther, to arrange with Captain Anderson to take us to the May in his converted fishing-vessel, the *Hilda Ross*. On a calm, sunny morning, accompanied by two friends, we sailed out of Anstruther harbour.

Once ashore Alasdair was in a fever to be off and take his photographs. Shedding the heavy clothing he had worn on the boat, he just disappeared. The incredible speed with which he moved always made accompanying him impossible. Now and then we would catch a glimpse of a swiftly moving figure on the skyline, or see it poised perilously on a cliff edge, bent over a tripod, oblivious to all and everything but the clouds and the sun and the view he was taking. The May was not to be the last island we were on together; but it and the Treshnish Isles were among the loveliest of all our island days.

In Orkney Alasdair had remarked to me that for the first time in his life he knew what it was like to feel tired. By the New Year, approaching illness had damped down the fire of his imagination. He knew then that he was fighting a losing battle; but by April 15th, 1970, the day of his death, apart from a factual checking of the last two paragraphs on Handa, this book was ready for the publisher, as were three other manuscripts.

My thanks are due to Ernest Benn Ltd., for permission to quote from *A Photographer's Notebook* by John Peterson; and to William Collins, Sons & Co. Ltd., for permission to quote from *Islands Going* by Robert Atkinson.

I am immensely grateful to Tom Steel, who with his wife, Dixie, shared with us that wonderful day on the May, for his help in arranging the lay-out of the photographs in *Islands by the Score*.

Patricia MacGregor

The King's Barn,
Odiham,
Hampshire.
July 22nd, 1970

Contents

Illustrations

ILLUSTRATIONS

PHOTOGRAPHIC ACKNOWLEDGEMENTS

By courtesy of British Insulated Callender's Cables for number 1. William Kay for number 15, 17, and 18. George Waterston for number 24. Campbell K. Finlay for number 26. Robert M. Adam for number 31.

Cape Wrath

Butt of Lewis
Port of Ness

Flannan Is.

Handa
Port of Tarbet

Stornaway

SUTHERLAND

N

LEWIS

Shiant
Isles

ROSS

HARRIS

N. UIST

BENBECULA

SKYE

S.
UIST

Canna

Barra
Vatersay
Pabbay
Berneray

Rum

INVERNESS

Castlebay
Sandray
Mingulay

Eigg

Muck

Coll

Tobermory

PERTH

Treshnish Is.

Ulva

Craignure

Staffa

ARGYLL

Iona

Fionaphort

LOCH
LOMOND

Balmaha

0   10   20   30
MILES

Jura

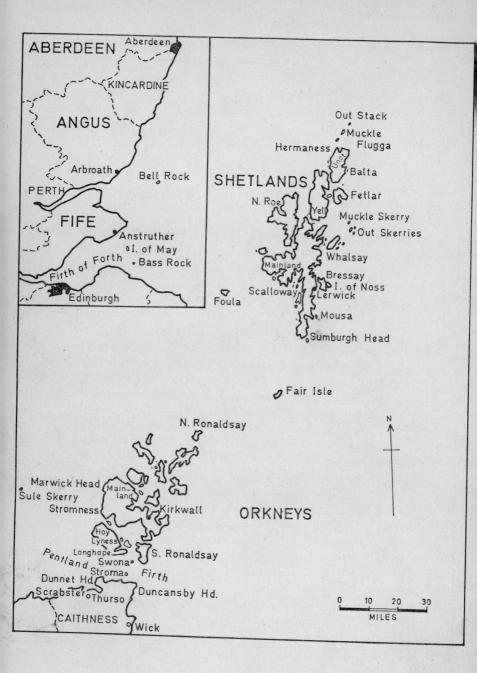

CHAPTER ONE

The Bell Rock
(where Ralph the Rover tore his hair)

NOBODY FAMILIAR with the extent to which the northern and western coasts of Scotland are dangerously beset by isles and islets, by sunken rocks and jagged skerries, many of them extremely isolated, is surprised that a country so comparatively small should have occupied a pre-eminent position in pharology. Its lighthouses, towering above tides surging ceaselessly round some of the world's most perilous sea-rocks, are numbered among the earliest and most historic. Nowhere can there be seen more impressive and enduring specimens of what lighthouse builders, in face of tremendous odds and repeated setbacks, and even dire adversity, achieved off the coasts of Scotland toward the close of the 18th century, throughout the 19th, and during the opening years of the 20th. This was mainly in response to the requirements of the world's rapidly expanding commercial shipping, particularly that between North America and Scandinavian ports, the choice of routes between them being some hundreds of miles shorter by way of the north of Britain than by the English Channel and the Straits of Dover.

Inseparable from the several achievements to which Scotland's lighthouses testify is the heroic surname of Stevenson, one of the worthiest in the long and colourful annals of the Scottish Capital. Successive generations of this family of civil engineers were associated with the Northern Lighthouse Board from 1786, the year it was instituted and assumed responsibility for lighting the perils of the Scottish seas, until the death in 1938 of David Alan Stevenson. The family's civil engineering business was to continue until 1952, however. To the Stevensons' renown in so

many fields must be added that of the most illustrious member
of the family—that of Robert Louis Stevenson, son of Thomas
Stevenson (1818-1887), of whom the former writes so lovingly in
his *Memories and Portraits*. In years more recent, the family's
literary tradition was maintained by the late D. E. Stevenson,
R.L.S.'s first cousin once removed, and also by D. Alan Stevenson, who retired from civil engineering in 1952 in order to
devote himself to what he has since been, namely, a technical
historian. Among the appointments the last mentioned has held
are that of engineer to the Clyde Lighthouse Trust and honorary
engineer to Scotland's Royal Burghs.

That eight members of the Stevenson family, in unbroken
succession, should have held the office of engineer to the Commissioners of Northern Lighthouses for over a century and a half,
and been responsible as such for the building and maintenance
of the numerous lighthouses situated along and at varying
distances from the Scottish coasts, is in itself remarkable. These
flashing lanterns, each displaying its own precise character, are
their enduring monument. It should be made clear, of course,
that, while the family practised civil engineering over a wide
field, it concerned itself chiefly with lighthouses, with beacons,
buoys, fog signals, harbour and river improvements in Britain,
and with the erection of lighthouses in countries as distant from
one another as are Newfoundland, Japan, and New Zea-
land.

Members of this family, throughout their careers, were
designers and consultants, *not* contractors. Their heroic under-
takings were initiated by Thomas Smith when, in 1786, he
became the Commissioners' first engineer, the office he occupied
until 1798. Thomas was succeeded by his son-in-law, Robert
Stevenson (1772-1850), whom he had taken into partnership
several years earlier. Thus Robert is recognised as the founder
of that famous Stevenson dynasty of civil engineers which provided the Commissioners of Northern Lighthouses with engineers
from 1798 until 1938, the year D. Alan Stevenson was succeeded
by John Oswald of the civil engineering firm of Gardner &
Oswald. Both of these partners I knew well in my early
life.

Robert Stevenson was responsible for a great number of lighthouses. His son, Alan, in that admirable *Memoir* of his father, puts the number at 27, while reminding us at the same time of his father's having invented the intermittent flashing light. Incidentally, Robert also designed the noble eastern road approach to Edinburgh—the ascending Regent Road route to Princes Street by way of Abbeyhill.

Up till about 1830, the Stevensons' business was carried on by Robert at Numbers 1 and 2, Baxter Place, his Edinburgh home, then situated near the top of Leith Walk. Its side entrance is still to be seen. For more than a century after 1830, the family, for business-office purposes, leased one or two floors of the Northern Lighthouse Board's premises at 84, George Street, an arrangement mutually satisfactory.

The Stevensons' renown began with the Commissioners' resolve to illumine the dangerous Inchcape Rock, lying in the Firth of Tay, a dozen miles from the mainland, in a sea area notorious for its gales. Professor A. C. O'Dell of Aberdeen University, in his contribution to a fairly recent publication on the physical background of the Orkneys and Shetlands, defines a gale as a wind exceeding, for a period of an hour or more, a velocity of 38 m.p.h.[1] Anemometer readings have shown at the Bell Rock gusts responsible for assaulting seas of a power and magnitude incomprehensible to anyone who has had no personal evidence of their fury, of their terrible dynamics. The average annual hours of gales at the Bell Rock is 255.[2]

In lighting the menacing Inchcape reef contemporaneously with a rapidly increasing North Sea shipping was the Stevensons' and the Commissioners' first joint, major undertaking. One recalls in connection with it Robert Southey's ballad relating how the Abbot of Aberbrothock (Arbroath) erected on it the bell which, when tolled by wind and wave, warned mariners sailing these waters of its perilous propinquity, how the pirate, Ralph the Rover, cut adrift the bell in order to ensure shipwrecks upon it, and how in the end the wicked Ralph himself perished precisely there:

[1] *The Northern Isles*, edited by Dr. F. T. Wainwright, and published posthumously by Messrs. Thomas Nelson in 1962. Wainwright, at the time of his death the previous year, was head of the department of Anglo-Saxon Studies in the University of St. Andrews.
[2] The average annual at the Butt of Lewis, however, is 378.

Sir Ralph the Rover tore his hair;
He curst himself in his despair;
The waves rush in on ev'ry side,
The ship is sinking 'neath the tide.

Southey's verses as published in 1909 in the Oxford edition of his poems are aptly prefaced by a quotation from Sir John Stoddart (1773-1856) apropos the tradition showing why the Bell Rock lighthouse was so named. 'By east the Isle of May, twelve miles from all land in the German seas, lyes a great hidden rock, called Inchcape, very dangerous for navigators, because it is overflowd everie tide. It is reported in olden times, upon the saide rock there was a bell, fixed upon a tree or timber, which rang continually, being moved by the sea, giving notice to the saylers of the danger. This bell or clocke was put there and maintained by the Abbot of Aberbrothock, and being taken down by a sea pirate, a yeare thereafter he perished upon the same rock, with ship and goodes, in the righteous judgement of God.' The treacherous reef upon which stands the Bell Rock lighthouse is 2,000 feet in length and roughly 100 in breadth. At high spring-tides it lies completely submerged to a depth of 16 feet: at the lowest ebb no more than 4 feet of its crest are exposed. So frequent during the closing years of the 18th century and the opening years of the 19th were shipwrecks on it that in 1806 authority for the construction upon it of a lighthouse was given. With the completion by this time of the Eddystone, John Smeaton and his colleagues had won *their* battle against the elements. Why shouldn't Robert Stevenson and his associates achieve at the Inchcape Rock something similar, even although it was fully recognised that the latter presented engineering problems demanding a considerably greater measure of skill and tenacity? Bear in mind that this enterprise was prior to the introduction even of steam navigation. Consequently, all communication between the reef and the mainland—the transference to and fro of men and materials—had to be maintained by rowing-boat, or by sailing-craft taking advantage of every favourable wind and tide. No simple wireless communications in those days; no regular weather forecasts. Of the dangers, difficulties, and disappointments all this involved, of the discomforts and perils endured by Robert Stevenson himself and the 28 men who inhabited the barrack perched on logs upon the rock like a pigeon-house, Robert bequeathed to us in his own account a moving memorial. Its 500 quarto pages he dedicated to Jane, his only daughter.

Preliminary work at the Rock began in 1807. The stone required—not all of it granite—was quarried at Aberdeen, at Mylnefield, near Dundee, and at Craigleith, now part of an expanding Edinburgh. Only on the calmest of summer days, and when the tide was at its lowest, could any work be carried out at the site. The most favourable conditions never allowed of more than five hours' uninterrupted work at a time, which explains why the building of the Bell Rock Lighthouse took five years. Its completion was the outstanding achievement of Robert Stevenson's life. 'All knew the difficulties of the erection of the Eddystone Lighthouse and the casualties to which that edifice had been liable,' runs a passage in his own account; 'and in comparing the two situations, it is generally remarked that the Eddystone was barely covered by the tide at *high water*, while the Bell Rock was barely uncovered at *low water*.'

Some idea of the force with which the waves strike obstacles as exposed to their fury as the Inchcape Rock may be had from that known as early as May, 1807 to have been exerted when, during the building of this lighthouse, six large blocks of stone landed on the reef were cast over a rising ledge a distance of more than a dozen paces, and an anchor weighing over a ton was thrown up upon it. Although the sea at a distance of a hundred yards all round this sunken skerry has a depth ranging from two to three fathoms at low water, the impelling force of the waves casts up on the reef from time to time boulders of more than thirty cubic feet and weighing over two tons. So familiar with such boulders did the Bell Rock's first lightkeepers become that they nicknamed them *travellers*. At such times as ground swells, unaided by any wind, hurl these boulders on the reef, the lighthouse itself, towering 120 feet above the sea, is smothered in foam and spray to its very top. As early as November, 27th, 1827, the spray is recorded as having risen 117 feet above its foundations. This represented a wave pressure of nearly three tons on the square foot. Records kept at the Bell Rock between the middle of September, 1844, and the end of March the following year show a wave pressure of 3,013 pounds on the square foot.

As work on this lighthouse proceeded, the Commissioners showed increasing interest in it. Indeed, from time to time some of them paid it visits of inspection. At a meeting held in the lighthouse itself on July 19th, 1824, Robert Stevenson's singular triumph was acknowledged when the Lord Advocate moved that a marble bust of him be placed in its library 'in testimony of the sense entertained by the Commissioners of his distinguished talent and indefatigable zeal in the erection of that lighthouse'. In due course Robert's bust by Samuel Joseph was installed there.

Among the earliest of the truly famous to land at the Bell Rock was Sir Walter Scott. Accompanying Robert Stevenson there in 1814 as one of the Commissioners, he inscribed in its album the lines he entitled *Pharos loquitur*:

> *Far in the bosom of the deep*
> *O'er these wild shelves my watch I keep*
> *A ruddy gem of changeful light*
> *Bound by the dusky brow of Night.*
> *The seaman bids my lustre hail*
> *And scorns to strike his timorous sail.*

When one reflects on the magnitude and recklessness of public expenditure nowadays, so often on needless enterprises designed anachronously to provide employment, one reads with amazement that the adoption of Robert Stevenson's plan for the erection on the Inchcape Rock of a stone lighthouse was delayed for some time because of the amount of money it would entail. Although Robert had estimated the cost at £42,000, the Lighthouse Board feared it might turn out to be as much as £50,000. 'To the very last,' as Robert recorded, 'the bankers were in doubt as to their security on the dues for so great and hazardous an undertaking.' At the end of the day, when at last the Bell Rock light began to function, it was found that this mighty monument had cost no more than £41,000.

In 1966, this celebrated light, hitherto operated on oil, was electrified when Messrs. Thomas Justice of Dundee completed for the Northern Lighthouse Board their installation of a new lighting and power system together with modern fire-alarm and intercom. devices. The change-over necessitated the introduction of 1,100 yards of British Insulated Callender's Cables Ltd, cables and a variety of accessories, as well as insulated telephone cables and also shorter cables for the three diesel generator sets. Thuswise the Bell Rock's red and white 730,000-candlepower signal was superseded by one of approximately 2,000,000.

* * *

Tribute to the dedication, discipline, and privation which the maintenance of any lighthouse service demands was paid in the televised version of a script on our Northern Lights written and produced by Finlay J. MacDonald, and transmitted from Glasgow in June, 1969. Among those who contributed to this moving documentary was Dan Mitchell. Dan is the fifth generation in succession of Mitchell keepers in this service. His great-great-great-grandfather was employed at the building of the Bell Rock Lighthouse. Upon its completion he went into the Lighthouse service. Dan, who entered it in 1943, at the age of 17, told us something of life on the island-station at the extreme tip of North Ronaldsay, most northerly of the Orkneys.

To those who might be inclined to regard paraffin as somewhat out-of-date, it may come as a surprise to learn that the new Ronaldsay Lighthouse is still paraffin-lit—lit on the vaporised paraffin principle, which has proved so very reliable. This light, in common with so many, is magnified to several thousand candlepower by huge lenses worked by that old, simple, clockwork device of weights and chains ensuring, as in a grandfather clock, an unbelievably accurate movement.

Although the Ronaldsay keepers are a community *per se*, they are at the same time part of that remote island's life. Their children are among the 17 attending that island's school.

The number of lighthouses operated by automatic or semi-automatic means is steadily increasing. Mains electrical power now reaches most of them. Whereas the automation of *lights* presents comparatively little difficulty, this as yet cannot be said of *foghorns*. Trinity House has been devoting much attention to automatic fog-signalling, and hopes that before long our fog-signals may be entirely controlled automatically. So it looks as though the routine duties of keepers at lone rock-stations may soon be obsolete.

One wonders how many readers of these pages have any conception of the rigorous discipline our lighthouse service involves. Do you know that, once the keeper has entered the light room to begin his 4-hour shift, under no circumstances must he leave it? Do you know that not for a moment during that shift must he read anything, or listen to the wireless? Do you know that, in order to ensure that the lens shall never cease revolving at its strictly regulated speed, the slowly descending weight that keeps it going must be wound right up to the top every half-hour? If it should be allowed to reach the floor, the light would not operate, and the keeper responsible would be dismissed instantly. Very rarely does a light fail to function. There is on record an occasion when one would have done so through a mechanical fault, had not the keepers kept the lens revolving for 17 long, wintry nights— by *hand*!

Sule Skerry for some reason comes to mind. On that guano-covered rock, 30 miles west of Orkney, stands the most isolated of Scotland's lighthouses. A gully provides its only landing-place. Into this, weather permitting, the stout motor-launch of the *Pole Star*, one of the four vessels servicing Scotland's 300 lighthouses, beacons and buoys, noses her way every three weeks to put ashore a relieving lightkeeper and take one off, having already dropped at some distance to seaward a holding anchor at the end of a strong rope enabling the *Pole Star's* mate to ease her safely shorewards, and immediately to withdraw her in the event of unforeseen trouble or of such mounting danger as would arise with a sudden change in the wind's direction and the menacing increase in the swell eternally surging round rocks of this kind. The anchored rope is the means whereby the launch is prevented from being dashed against the rocks. It also provides the means whereby it may be hauled back to comparative safety should her engine fail. No time in such circumstances even for the exchange of formal greetings and politenesses!

Immediately the launch has left, the tackle at the landing-place is again made secure against the most violent of the ocean's onslaughts. Any damage to it, whether through human carelessness or storm, would render even more hazardous the next relief. Instantly the stores landed at Sule Skerry are loaded on to a bogie which is hauled by a cable fixed to a petrol-driven drum up to the lighthouse—very different from the years when, arduously and dangerously, the keepers had to carry everything up and down on their backs.

When the relief is not expected because of bad weather, and the three-weeks bread supply previously landed is consumed, the keepers can always bake what they require. Furthermore, as Finlay MacDonald reminded us, all rock-stations such as Sule Skerry are furnished with coffin boards in order to cope with the worst that may befall.

Most major lighthouses now have foghorns and radio navigational aids. The three men on each of them keep meteorological records. These they transmit to the meteorological office at Wick, whence they find their way into our weather forecasts.

There are still 72 manned lighthouses round the Scottish coasts. They and all the buoys and beacons, apart from those in the estuaries of the Clyde, Forth, and Tay, are under the Commissioners' authority. The Commissioners comprise Scotland's Lord Advocate and Solicitor-General, all her sheriffs-principal, and the provosts of several Scottish cities and towns intimately involved in maritime matters.

CHAPTER TWO

Isle of May

++

PARTICULARLY OVER the school-children of Fife and the Lothians, by which counties they are shared, the islands of the Firth of Forth have always exercised a fascination. During my own school-days in Edinburgh, three of its baker's dozen were my special concern—Inchkeith, because of its boasted impregnability as a naval fortress; the Bass Rock, on account of the popular cruises circumnavigating it during the summer and autumn months, and of its being so exclusively the territory of the solan goose as to explain the scientific significance of *bassana*; and the more distant Isle of May, accessible by small craft at suitable tides from the Fifeshire fishing-ports of Crail, St. Monance, Pittenweem, and Anstruther. At low tide, the harbours of these ancient burghs of the East Neuk dry out, as do also Kirk Haven and the Altarstanes, the two recognised landing-places on the May itself. Usually the more suitable and, consequently, the more frequently used is Kirk Haven, situated on the east side of that island and accessible in reasonably good weather by rounding North Ness and its north-eastern extremity, when the wind and the swell it accentuates are from the west. When both are from the east, a landing is nearly always possible at the Altarstanes, on the west side. The latter is less advantageous, however, because it provides no mooring at which a boat can be left unattended. This necessitates either its immediate return to the mainland, or its anchoring not too conveniently offshore. Kirk Haven, on the other hand, provides a little sand-bottomed harbour at which those intent on remaining for any length of time on the island may tie up with safety.

The Isle of May lies well down the Forth estuary, between five and six miles south-east from Anstruther, and roughly twelve

north-eastward from North Berwick, on the opposite shore. Its
southeast-northwest axis measures under a mile, just thrice its
maximum width. Composed entirely of greenstone, it has an area
of about 140 acres, roughly a ninth of which is its uninterrupted,
rocky coastline, a considerable additional area of which, due to an
average tidal rise and fall of 13 or 14 feet, lies exposed at low
water. This coastline is penetrated by several gullies and
caves. Immediately offshore stands a number of stacks, slowly
being worn away by the unceasing marine erosion that, in the
main, created them. Nowhere except at the west and south-west
does the island possess anything which could be regarded as
cliffs. At the centre, where it reaches its maximum height at 168
feet, is the Tower, the name given to its present lighthouse,
accessible both from Kirk Haven and from the Altarstanes by a
road not too rough for vehicles of the land-rover class. Directly
north of the Tower, and at a distance from it of no more than 500
yards, lies that whitewashed building on the low, eastern shore
known as the Low Light. This subsidiary lighthouse of earlier
years is now the headquarters of the Isle of May bird observatory,
established in 1934 by members of the Midlothian Ornithological
Club primarily at the instigation of my friend, George Waterston.
The Club occupied, at the outset, the old coastguard lookout. In
1946, when its activities, interrupted by the Second World War,
were resumed, the Commissioners of Northern Lighthouses, to
whom the Isle of May belongs, allowed it to shift its quarters
to the more commodious Low Light, where they have been ever
since. The two lighthouses' lying in a north-south line renders
extremely easy not only the estimating of wind direction, but also
the direction of approach and of departure of migrating
birds.

The May is transversed by several faults, three of which the sea
has so eroded as to divide it at high water into four parts. The
northernmost, flat and barren, is North Ness. Then comes Rona,
that part on which stands the North Horn. Beyond this, south-
eastward, lies the bulk of the May. Lastly comes the skerry known
as the Maiden Rock. Deeper than any of these faults, and
existing today as a steep-sided valley, is the Black Heugh, which
virtually bisects the island. The damming of this valley's west
end formed the small, freshwater loch or reservoir, to the east of

which are situated the principal lightkeeper's residence and the lighthouse's engine-house.

The lightkeepers and their families are the only established human residents the May now knows. Each family has its own house and garden. Each as a rule keeps a few hens, perhaps a dog, and usually a goat or two. Sometimes this treeless isle carries a few sheep, finding sustenance on such vegetation as the numerous rabbits may not have consumed or rendered inedible. This vegetation consists mainly of a sward of red fescue which, near the island's modestly cliffy edges, tends to give way to thriving clusters of sea-pink and sea-campion growing on a shallow soil composed for the most part of their own natural compost. Except where, in a few cracks and crevices, tufts of fescue and other grasses retain a precarious root-hold, the bare rock comprising most of the surface of the North Ness is completely devoid of vegetation. Nettles, hemlock, and burdock are found there in limited quantities. Much more luxuriant are parts of Rona, especially where, in summer, an adequate depth of seabirds' droppings, accumulated over the centuries, supports a healthy growth of orache and chickweed. How limited from the serious botanist's point of view is this list of mine can be seen from the meticulous classification of the May's flora appearing in W. J. Eggeling's *The Isle of May*, one of the best written, best illustrated, most orderly, scholarly, and beautifully produced books I have ever possessed.[1] For an island so limited in size, this bibliography is immense, covering its every aspect—geology, archaeology, history, ecclesiology, toponomy, pharology, and natural history, with particular emphasis on its ornithology.

With a range of roughly 20° Fahrenheit between summer and winter temperatures, the May enjoys what meteorologists describe as an equable climate. Its rainfall, distributed fairly evenly over the island and throughout the year, averages 22 inches. Although this, in relation to the smallness of the area on which it falls, is not sufficient to maintain anything that could be regarded as running water, it does produce a number of pools, nearly all of which dry out entirely during the summer months. Here and there, on the other hand, small patches of ground are maintained in a moist

[1] First published in 1960 by Oliver & Boyd, at Tweeddale Court, Edinburgh.

condition by the tiniest of surface springs. These, of course, are quite insignificant when compared with the island's five wells, all of which, at one time or another, were resorted to regularly, as is shown by their having been symmetrically stone-lined. Only that known as St. Andrew's Well now provides drinkable water. It supplies today the needs of the bird observatory, so close to which it lies. The lighthouse's drinking-water requirements are met entirely by the water landed at regular intervals from the Lighthouse Commissioners' vessel based on Granton, on the opposite side of the Firth.

That the drinking-water situation on the May was little better a century ago is shown by T. S. Muir in that masterpiece of his, *Ecclesiological Notes on Some of the Islands of Scotland*.[1] Muir managed to make a couple of trips to the May in 1868 in order, primarily, to examine what then remained of its priory, inferring from the character of two windows in its west wall that it was 13th-century, and noting a number of internal additions to the original structure without in any way injuring it—the insertion of a large press or locker in the west wall, for example, and what was then the fairly recent interpolation of an oven in the bottom of the south window. After examining the island from a holiday viewpoint, and regarding as a serious limitation the absence of sandy beaches and nooks suitable for bathing, Muir continues: 'What I allude to is an equal if not greater destitution in the article of water—not *aqua marina*, of course, but *aqua*, of the sort your comforting and sleep-compelling nightcap is made up with! Not that boon Nature has denied wells to this bit of her territory. There are some four or five in the May, which is pretty fair allowance, I take it, considering how parsimoniously many other deserving islands I could name have been treated; but all of them are dried up at present except one or two, and these are so intermixed with impurities that no use can be made of them. To obviate the inconvenience arising from this local infirmity, the lighthouse people are obliged to procure water by periodical supplies brought over from Crail. Yet, after all, this does not much mend the matter; for the supply thus obtained, though excellent when taken from the fountain, acquires during transmission such an offensive taste of the casks, that the islanders, notwithstanding

---

[1] Published in Edinburgh by David Douglas in 1885.

their endeavours to accommodate themselves to it, never drink it but under a kind of protest.'

Muir then proceeds to furnish us with a lengthy footnote naming and supplying details of the May's five wells—the Lady's Well, the Pilgrims' Well, St. John's Well, St. Andrew's Well, and the Sheep's Well. Details of each he had obtained from Joseph Agnew, the May's principal lightkeeper. That there were cattle on the island at this time is shown in what Agnew had to say about the Pilgrims' Well, and St. John's Well. 'During all the drought of this summer [1868] we pumped water out of this well to supply our cattle', he added in connection with the latter. There were on the May at that time three milch cows and some sixty sheep. As the water of St. Andrew's Well was still the best on the island, the lightkeepers resorted to it for culinary purposes when their own supply, then brought over from Crail, ran short.

The expression of the elements to which the Isle of May is thoroughly inured is fog. It receives its due share of what, as a schoolboy at Edinburgh, I soon got to know as the East Coast Haar. The other day, in mentioning to Dr. Harry Lillie, the noted physician and naturalist, my recent visit to the May and the touch of fog shrouding much of it when I stepped ashore, he recalled something which he remembered his father's having told him in boyhood at Crail. At the turn of the century Mr. Lillie, in response to emergency calls from the May's lightkeepers, often sailed the local doctor, Rutherford Dow, over to the island. While crossing together on one occasion, a dense bank of fog descended so suddenly as to have necessitated their steering as best they could by the intermittent boom of the island's foghorn. When halfway across, the horn, as they thought, ceased. Continuing on their course, they eventually located the landing-place by the noise of the waves on the rocks. On reaching the lighthouse, they immediately asked the keepers why they had stopped the foghorn. 'Stopped the foghorn!' they replied in astonishment. 'It's going now as it's been going for two days!'

The horn's sound-waves could not penetrate audibly the fog where the angle dropped approximately midway between Crail and the island. It was as the result of this experience of theirs that the present alternating high and low horn notes were introduced at the May. For the same reason, two such notes were first installed in Canadian Pacific steamships and railway locomotives.

Although already firmly established in ancient lore and legend, the May's recorded history does not appear to have begun until the foundation of its priory about the middle of the 12th century, during the reign of Scotland's David I. Some three centuries earlier, according to legend, the zealous missionary, St. Adrian, and a number of his followers landed on the May in the course of their endeavours to Christianise the Pictish inhabitants of that part of Scotland off which it lies. They were getting on very well with their missionary work in this region when, according to Andrew Wyntoun's *Orygynale Cronykil of Scotland,* the Danes arrived in great, murderous numbers and slew them all:

> *And apon Haly Thurysday*
> *Saynt Adriane thai slwe in May,*
> *Wyth mony off hys cumpany.*

Scholars give the date of Adrian's martyrdom as Thursday, March, 4th, 875. On that day of the year—the Day of St. Adrian the Martyr—the inhabitants of Pittenweem, for centuries thereafter, held one of their two public markets.

The May found its first niche in history when, about the middle of the 12th century, King David, the 'Sair Sanct tae the Croun', renowned for his having founded abbeys and sees in Scotland, made to Reading Abbey a gift of its fishings. These proved so lucrative at this time that the May received regularly in its harbours not only Scottish fishermen, but also fishermen from England and from the coasts of France and Belgium. With this endowment would have originated the island's earliest monastic recognition even if at the outset this amounted to no more than a cell for its administration. In the course of a few years the May could boast at least something of the priory substantially endowed alike by kings and noblemen during the latter half of the 12th century and the first few decades of the 13th. This priory remained in the possession of the Reading monks until 1288, when their abbot, Robert de Burghgate, sold it somewhat arbitrarily to William Fraser, Bishop of St. Andrews. In 1305, after some years of confusion, the priory once more reverted to Reading Abbey, dependent upon which it continued until the defeat of

the English at Bannockburn in 1314. Thereafter it again came
under the jurisdiction of St. Andrews. That it was completely so
by 1318 is seen in the deed of gift dated July, 1st, of that year,
whereby Bishop William of St. Andrews, with the concurrence
of Martin, Prior of the May, made over to the canons of its
monastery the pension of 16 merks which hitherto the former
had remitted to Reading.

Exactly when the monks quitted the May is uncertain. One
does know, however, that until the middle of the 16th century
the island remained in their possession, and that they continued
to derive from it and from the waters surrounding it such harvests
as they could. Scarcely had they abandoned it because of its
unsuitability for permanent residence when English raiders landed
to pillage and to destroy much that, today, would have been of
considerable archaeological interest. Before levelling the priory's
walls, they robbed it of anything deemed worthy of removal.

Upon this priory's site, and at a date one cannot give with any
degree of precision, was built the chapel or hermitage commemorating St. Adrian and other saints traditionally claimed to
have been buried on the May. In no time the island was to become
a noted objective of pilgrimage, due largely to its increasing
reputation as a place where miracles were performed. In the late
15th-century Breviary of Aberdeen, not the least satisfactory of
these is seen to have concerned the women who, deploring their
childlessness, arrived at the May to drink of its Pilgrims' Well.
'Barren women, especially,' as the Breviary puts it, 'coming in
the hope of thereby becoming fruitful, were not disappointed.'

By this time St. Adrian's Chapel was recognised as a place of
retreat for Scottish Royalty. In 1513, shortly before James IV met
disaster at Flodden, he erected the lands of Admiral Sir Andrew
Wood of Largo into a free barony in order that 'he, being skilful in
pylotting,' should conduct him and his queen to St. Adrian's
Chapel. Nearly a quarter of a century earlier, Sir Andrew had won
royal favour because of the skill with which his ships, the
Mayflower and the *Yellow Carvel*, had routed off the May a
substantial English naval force, and had brought captive into Leith
no fewer than five of its ships. Toward the close of the 15th
century and during the opening years of the 16th, the May was
visited regularly by other prominent members of Scottish society

anxious to avail themselves of such curative properties as were believed to reside in its principal well, and in St. Adrian's Chapel. Meanwhile, the island's use as a base for smugglers and for pirates plundering ships in this locality grew apace, its caverns affording ample storage for what had been taken from vessels intercepted in neighbouring waters throughout the 16th and 17th centuries, and during the early years of the 18th.

As early as 1577, accounts of piracies committed by those who lurked there or thereabouts began to find their way into the Register of the Privy Council of Scotland, since they interfered not only with trading vessels, but also with the local fishermen. Some years previously a French ship of war frequenting May waters had been named as a participant in this piratical traffic. Pilgrims to the island, apprehensive of molestation, now became chary about visiting it. The monks consequently began to find their retention of it increasingly unprofitable and precarious. Even the revenue from its rabbits had ceased by this time, an English force having landed and destroyed their warrens. So, in 1549, Prior Roull feued it to Patrick Learmonth of Dairsie, provost of St. Andrews. Two years later it passed from Patrick to Andrew Balfour of Mount-quhanie. In 1558 it came into the possession of John Forret of Fyngask on the understanding that, if war involving Scotland should break out, he should not be held liable for feu duty while hostilities lasted.

About 1570, when the May already had become part of the Fifeshire ecclesiastical parish of Anstruther Wester, a certain Allan Lamont became feuar; but for how long I cannot say. From Allan, as Eggeling records, it was acquired by the Crail laird, Cunningham of Barns, in whose family it remained for almost a century.

In 1639 the Crail laird, Alexander Cunningham of Barns, erected on the May its first beacon, to which we shall refer later. In 1714 the Cunningham family was succeeded as feuars by the Scots of Scotstarvet. From them it passed soon afterwards to the Balcomie family. In 1814 the Commissioners of Northern Lighthouses purchased it from the Duchess of Portland, who had inherited it from her father, Major-General John Scott of Balcomie. The Commissioners still own it. In 1956, by agreement between the Nature Conservancy and the Commissioners, the

May became what it is now, namely, a National Nature Reserve.

To the emphasis placed on the May's ornithological importance may be attributed the sparse archaeological attention it has received in recent years. Bird-watchers would hardly welcome in their midst the noisy activities of those who might be engaged in digging on this same, circumscribed field. Consequently, we know comparatively little either of the ruined Church of its Priory or of St. Adrian's ruined Chapel. For over a century they have been neglected. In 1868, after the Ministry of Works had done some repointing there, and also carried out a little excavation along one or two of their walls, the archaeologist, John Stuart, expressed in his *Records of the Priory of the Isle of May* the hope that the Chapel might receive treatment to ensure for many years the preservation of a memorial closely associated with the early history of this locality. Similar hope was expressed in 1933 in the Eleventh Report of the Royal Commission on Ancient Monuments in Scotland, a Stationery Office publication drawing attention to the sorry condition of St. Adrian's Chapel and its mediaeval lancet windows, and emphasising the need for protection against further deterioration. Stuart mentions, *inter alia,* that there was within the Chapel 'a stone coffin with covered head, and a bottom, formed of one stone. The rest was probably composed of separate slabs'. During the war years, 1939 to 1945, this relic, thought to have been 13th-century, vanished from the May. In 1958, St. Adrian's Chapel, and also the island's old Beacon (to which we shall refer) were scheduled as Ancient Monuments. But that is more than a decade ago; and as yet, so far as I am aware, neither restoration nor preservation work has been undertaken.

When wandering over the May's bleak acres, one has difficulty in believing that, in the midst of all the piracy and smuggling its situation facilitated, there existed upon it, throughout the 16th century and the opening decades of the 17th, a village, traces of which, a little to the south-east of its ruined Chapel, were clearly distinguishable up to the end of last century, but which today would require to be unearthed. That in the village's immediate proximity lay a burial-ground of some sort is suggested by the solitary and broken headstone marking the grave of the last villager to die there and be buried there:

Here lyes John Wishart husband to
Euphan Horsbrough who lived on the
island of May who died in March the 3
1730 aged 45.

Wars have usually brought to the May a particular significance because of its strategical position in relation to the Forth's mighty estuary. During the Jacobite Rebellion of 1715, when the Earl of Mar was busy transferring a force from Fife to the Lothians, a contingent of 300 men, under the young Earl of Strathmore, was driven to seek refuge there, and had to remain there for eight days before he was able to withdraw in darkness to the Fife coast.

Although throughout the wars of our own century the May was garrisoned, it never experienced anything in the nature of an attack. Deadly conflict constantly rent the seas and skies at no great distance from it, however.

The comprehensive place-names map of the May reproduced as endpapers in Eggeling's book, as well as the several closely printed pages on place-names inserted as an appendix, enshrine much of legendary, traditionary, archaeological, historical, and topographical interest. Because of a unique combination of such, few islands as small, infertile, and at times as inaccessible and even forbidding, have merited and been accorded a toponymy as precise, intimate, and picturesque. Eggeling lists more than 160 place-names, supplementing several of them with appropriate notes on their origin—a veritable gazetteer of an isle of but 140 none too hospitable acres! His earliest entries, many of considerable antiquity, as many associated with the monks and their profitable fishing charters in mediaeval times, derive from place-names in Sir Robert Sibbald's *History, Ancient and Modern, of the Sheriffdoms of Fife and Kinross,* published in 1710 at Cupar, the county-town of Fife. Then come place-names not mentioned by Sibbald, but appearing on the original 6-inch ordnance map of 1854. The newest place-names now recognised are, of course, those which have been added since the establishment of the bird observatory on May in 1934. Doubtless, further additions have accrued since 1954 from the island's National Nature Reserve status.

The May's niche in pharology is of interest. As may be seen
in the Register of the Privy Council, Alexander Cunningham in
January, 1631, made application to that body to be allowed to
instal on the island a light for the guidance of the ships then
frequenting the Firth of Forth in ever-increasing numbers. Acting
on a remit from Charles I, but without showing much enthusiasm
for Cunningham's proposal, the Council directed that letters be
addressed to the provosts and bailies of Edinburgh, Dundee, St.
Andrews, Anstruther, Crail, Pittenweem, Kirkcaldy, Dysart, and
Burntisland, requiring each to send a representative to advise
'anent ane propositioun made to the Kingis Majestie for erecting
of lichts upon the Isle of May, as ane thing thocht to be maist
necessarie and expedient for the saulfetie of shippes arryvying in
the Firth'. Having heard what these representatives, and also
the skippers of a number of fishing and other vessels constantly
plying these waters, had to say in favour of Cunningham's pro-
ject, the Lords of Secret Council rejected his application. They
saw no reason why any duty should be imposed for the maintenance of a beacon such as he had in mind. This setback did not
discourage Cunningham. On the contrary, it inspired him to
canvass the support of all the fishing and seafaring communities of
these coasts. That he did so with the result he desired is seen
in the King's granting his application in 1636, five years later.
In April of that year, the Council duly authorised Cunningham
and his son, John, and James Maxwell of Innerwick to levy on
ships using the Firth an impost such as would allow of their in-
stalling and maintaining a beacon-light on the May at an annual
rental of one thousand pounds 'in coin of this realm,' or £84
sterling. In 1641, the Scots Parliament ratified this authorisation.
Four years later, the Privy Council confirmed the tenure of Sir
John Cunningham, acknowledging in picturesque terms the 'true
and thankeful service to his hieness and to the estates of this
kingdom doune be his Majesty's Lovitt Johne Cunnynghame of
Barnes in bigging and erecting upon the yle of Maii belonging
to his lyand in the mouth of the firth of ane lighthouse and
keeping and menteaneing of light thairine continowallie in the
night tyme for the saiftie and directing of sailleres in thair in-
comeing and outgoeing of the said firth in the darke nightes upon
his great charges and expenss'.

The beacon itself—the simplest thing imaginable—was just a large, legged grate or brazier mounted on the flat roof of a square, two-storeyed tower, its roof and embattled parapet reached from within by a spiral staircase up which coal in great quantities was carried. Not until after 1786, the year that the Northern Light-house Commissioners came into existence, was an external windlass arrangement of rope and pulley erected at one side of the original tower to facilitate the raising to the roof of the pans of coal re-quired. Sibbald describes memorably the original tower and its function—'a tower forty feet high, vaulted to the top and covered with flagstones whereon, all the years over, there burned in the night-time a fire of coals for a light; for which the masters of ships are obliged to pay for each tun two shillings. This sheweth light to all the ships coming out of the Firth of Forth and Tay, and to all places betwixt St. Ebb's-head and Redcastle near Montrose.'

The single-handed maintenance for nearly a century and a half of the May's primitive beacon-light—the first of Scotland's lighthouses, and for a long time her only one—must have entailed for each keeper in turn a considerable amount of arduous labour, if not also of danger from time to time. In the short nights of the northern summer, the beacon consumed under a ton of coal: during the long, wild, wet nights of winter it con-sumed thrice that quantity. Detached from it, quite early in its career, Alexander Cunningham erected for the keeper, in addition to such room within the tower as he was able to adapt to his own purposes, 'a convenient house, with accommodation for a family'. But tragedy lurked not far off. The beacon had been operating no time when Cunningham, returning home to Crail during a storm, was drowned. As this was at a period when witch-hunting was an earnest pursuit among the fishing communities scattered along the Fifeshire coast, his death was attributed to the machinations of at least some of the women who, at Pittenweem, a few years later, had confessed to their participating in witchcraft and were duly burned.

Strenuous as was the maintenance of the May light from the outset, not until the closing years of the 18th century did tragedy of the grimmest kind overtake those responsible for it. In 1791 the keeper, George Anderson, his wife, and five of their six children were found inside the Beacon suffocated by fumes from smouldering ashes. Only the Andersons' youngest infant, a recently appointed second keeper, and a junior assistant managed to reach fresh air soon enough to ensure their survival. This fatality necessitated, in the interest of the May's lightkeepers and of any children they might have, better housing arrangements.

To record all of interest anent this original lighthouse throughout its active existence would occupy more space than now remains at our disposal. Suffice it to say that during the year, 1800, nearly half a million tons of shipping paid dues in respect of its services. In 1814, for the sum of £60,000, the Isle of May, together with its light dues, was acquired from the Duke of Portland by the Northern Lighthouse Board. Forthwith began work on the erection of the Tower, the present lighthouse on the May which replaced it eighteen months later. Throughout the night of January, 31st, 1816, the old Beacon performed its last duty. The following evening, there shone from the Tower the Light—a stationery oil light, fitted with a lamp and reflectors, standing 168 feet above the sea and visible for about 20 miles—that took over from it.

For what survives of the old Beacon we have to thank Sir Walter Scott. On July, 29th, 1814, shortly after the Lighthouse Board's purchase of the May, Scott landed there as one of the Commissioners then on their annual cruise of inspection of Scotland's lighthouses. In his diary for that date runs the following entry: 'Reached the Isle of May in the evening, went ashore, and saw the light—an old tower, and much in the form of a border keep, with a beacon-grate on the top.' When Robert Stevenson, the Board's engineer (R.L.S.'s grandfather) mentioned to Scott that he proposed razing the old Tower as soon as the new lighthouse was functioning, Scott immediately suggested 'ruining' it à la picturesque. That is to say, demolishing it partly. The Board, with some modifications, adopted Scott's suggestion, lowering it by about 20 feet to present the squat, conspicuous, single-storeyed, whitewashed, 20-foot square, 20-foot high building we see there today. Its roof underwent some alterations later. The battlements were added in 1886, so that to some small extent it might resemble architecturally the lighthouse buildings which replaced it. Built of local rubble with freestone quoins, its interior was converted into a guardroom for the convenience of the Forth's pilots and fishermen. Its entrance and its only window face south. Over the former is a largely empty panel-space surmounted by a cornice. The missing panel, as Eggeling records, has been replaced by part of a pediment bearing a sun in glory and the date, 1636, and also the two portions of an angle water-spout in the form of a lion supporting a shield. The Beacon, that proud objective for so many of us Edinburgh schoolboys, when summer holidaying at the East Neuk was crowned by a day's exploring of the May, is used today as a store for the Tower, the lighthouse that superseded it. In 1958 it shared with the Chapel the distinction of being scheduled as an Ancient Monument. This lighthouse, a two-storeyed building of grey stone providing accommodation for three keepers and their families, is surmounted by a square tower situated over the central part of the building's front. Surmounting the Tower in turn is the domed lantern-room standing 80 feet above the building's foundations, and 240 above the sea. In 1843 this lighthouse's beam, originally fixed, was replaced by a revolving light which was still worked by oil. This functioned until the installation in 1886 of electrical equipment rendered the May's light the first electrically operated in Scotland. On the ground that there was not a more important lighthouse station in Scotland, 'whether considered as a landfall, as a light for the guidance of the extensive important trade of the neighbouring coast, or as a light to lead into the refuge of the Forth,' the Board of Trade, three years earlier, had authorised the expenditure of the £16,000 this change-over would entail. Much of this went in the installation of engines, in the erection in what is known as Fluke Street of ancillary buildings, and in the conversion of the little loch lying in the Black Heugh into a reservoir for water required by the engine-house.

This introduction of electricity at the May meant a substantial increase in maintenance cost. To begin with, it necessitated the retention on the island of no fewer than seven keepers. Toward this cost an Order in Council of 1886 authorised the collection from passenger and cargo ships operating in local waters and benefiting from it, of a light due of $\frac{1}{8}$th of a penny per ton. The dues of such ships sailing to or arriving from foreign ports were fixed at a penny per ton.

The next change-over at the May, necessitated by the high

expenditure entailed in maintaining this electrical installation, was made in 1924, when a single incandescent mantle, burning vaporised oil, and many times magnified, became the luminant. This made possible the reduction of the operating staff to four. Besides attending to the radio beacons and to numerous other safety appliances, this staff sees to its two foghorn compressors. The South Horn, installed in 1886, was reconstructed in 1918. The North Horn was installed just prior to the outbreak of the Second World War.[1]

The substitution of the present system of illumination at the May was one of significance not only to mariners and pharologists, but also to ornithologists. Throughout the years that the light was electrically operated, enormous numbers of birds dashed themselves to death against its lantern's glass prisms. After a foggy night in the migration season, scores lay dead or dying on the flat roofs immediately beneath the lantern-tower. As many as 400 birds, representing 30 species, perished in this way during one short, July night.

For all its lighthouse improvements, for all the ominous groanings of its foghorns over the years, the Isle of May and its more immediate waters have remained the scene of shipwreck, although, on the whole, with remarkably little loss of life. In the last hundred years, 40 vessels are recorded as having gone on the May. Half of these became total wrecks: half managed eventually to get off, either unassisted, or with the help of a tug or two. Evidence of the former, sometimes quite substantial, lies strewn upon its rocks. In the files of the Fife and Lothian newspapers are preserved more accurate details of these disasters than they would be accorded nowadays.

One of the disasters traditionally recounted along the Fife coast during my Lowland schooldays was that which occurred in 1837, roughly three-quarters of a century earlier. I remember vividly an afternoon's shore-wandering at Crail when an old fisherman plied me with the grimmest details of it. Relatives of his own, he told

[1] Some idea of the precision all this entails may be had from a footnote with which W. J. Eggeling furnishes page 43 of his thoroughly informative book on the May. The South Horn, he tells us, emits four blasts of the same pitch in quick succession every $2\frac{1}{4}$ minutes, the duration of each blast being $2\frac{1}{2}$ seconds. The North Horn gives a blast of 7 seconds every $2\frac{1}{4}$ minutes. The two signals never sound together: they begin to blare $67\frac{1}{2}$ seconds after each other. What precision!

me, were among the fourteen who lost their lives. 'It happened on July, 1st, 1837—a lovely, summery Saturday.' The occasion was the annual excursion to the May from Fifeshire's coastal towns and villages. Half a dozen boats set out in the usual way from Anstruther in the most promising conditions. One of them, the *Johns*, a boat 36 feet in length and 12 in width, sought to land her 65 passengers at the Kirk Haven, instead of at the Altarstanes, which seemed the more propitious. Somehow or other, and with the tragic result mentioned, she struck a rock. Subsequently, her master was charged with culpable homicide. He was accused of having overcrowded perilously a boat of the *Johns's* dimensions, of having made for the Kirk Haven when he should have gone round to the Altarstanes, and of having had the use of no more than 4 oars when he ought to have had at his disposal 6 or 8. Although he was acquitted on the evidence given by those in charge of the more fortunate boats, this annual excursion was discontinued.

The next notable local mishap at sea was that in which the *Windsor Castle* was involved on the 1st of October, 1844, when, with 200 passengers aboard, she struck, without loss of life, the reef known as the North Carr Rocks, lying a mile north-east of Fife Ness and seven due north of the May, warning of its position now given by a lightship, but at that time by a beacon buoy. What made the more annoying this mishap was its having occurred so soon after the Low Light, the small lighthouse on the east side of the May, was first brought into use. This stationary light was designed primarily to indicate to shipping the Carr Rocks' position while, between 1843 and 1844, the Tower's fixed light was being replaced by a revolving lantern. Not until toward the end of the 19th century was the Low Light superseded by the North Carr lightship. During the Second World War the Low Light was occupied by men of the Royal Observer Corps. In 1946 it became the headquarters of the May's bird observatory.

As fortunate as had been the crew of the *Windsor Castle* 28 years earlier, was the crew of the *Matagarda*, the three-masted schooner of 150 tons that in April, 1872, ran ashore on the May, at the North Ness. The miraculous escape of her crew of six, as related by Joseph Agnew, the principal lightkeeper, makes dramatic reading. Agnew had but reached the scene when a terrific wave lifted the schooner to result in her crashing on the rocks with such violence that her keel was thrust up through her deck. The two members of her crew, swept away when attempting to get ashore, clung to the rocks until the lightkeepers were able to throw out to them the line by which they were retrieved. Agnew ordered the other four to remain aboard the schooner. This they did, with the result that some hours later, on a falling tide, they were got safely ashore.

Bringing the May's shipwrecks more up to date, one recalls the three cited by Eggeling. First of these was the *Mars,* a Latvian cargo-boat of 540 tons. She went aground on the North Ness on a foggy morning in May 1936, when on her way to load coal at Methil. There was no loss of life, since she was not abandoned until the Anstruther lifeboat had taken off her captain and her crew of 13. In April of the following year, that same lifeboat saved the total of 66 aboard the *Island,* a Danish passenger and cargo vessel of 1,774 tons that early one morning, in a dense fog, while plying between Copenhagen and Iceland by way of Leith, struck the May near Colm's Hole, an inlet on its east side.

As recently as December, 1959, this same Anstruther lifeboat picked up and brought to safety from their rafts the crew of the *Thomas L. Devlin.* Late one Sunday evening just before Christmas, this Leith trawler was returning to port with a heavy catch when she ran ashore at the North Haven. Some weeks later she slipped seaward to sink out of sight.

Although everywhere and in every field so much scientific investigation has been carried out in recent years, to few of our lesser British islands has such meticulous attention been devoted as to the Isle of May. Learned papers on every aspect of the exposed greenstone sill, of which it is composed, are numerous. Geologists have described it and such adjacent skerries as the Middens and the Maidens in the minutest detail. The ample evidence it provides, for instance, of the submergence and emergence of land has been recorded with commendable accuracy. In the locality of the Pilgrims' Haven, near its south-western extremity, proof of the latter exists in its raised beaches. These correspond with similar ancient terracing clearly discernible along the shores of the Firth of Forth. As easily found on the May as on the Fifeshire mainland is evidence of the Ice Age. Strange as

The Bell Rock Lighthouse in 1966, the year it was electrified to give approximately 2,000,000 candlepower. Hitherto it had operated on oil at 730,000 candlepower

Drawing of Samuel Joseph's marble bust of Robert Stevenson in the Bell Rock Library

ISLE OF MAY. Kirk Haven and the South Horn at low water

ISLE OF MAY. The Low Light from the Burrain, with the North Horn beyond. In the distance the Fife coast between Crail and Anstruther

it may seem, marine erosion and the epigene weathering of its rocks have contributed less to its topography than has glaciation.

What may come as a surprise to most readers is the fact that the rocks, cliffs, and offshore skerries of an island so small, an island nowhere attaining an altitude of any significance, should provide rockers with a sufficient number of adventurous climbs as to have added appreciably to Scotland's rock-climbing literature. Great altitudes, of course, do not necessarily furnish great climbs in the rock-climbing sense. A rock-face but a couple of dozen feet in height may confront the climber with more perilous problems than would several dozen demanding little or no foot-hold and hand-hold as understood in mountaineering and cliff-climbing circles. Vertical wall-faces, such as those in the vicinity of the Pilgrims' Haven, should be tackled only by the experienced rocker. The successful ascent of the Pilgrim and the Angel, its inlet's two stacks, rising dramatically quite close inshore, represents no mean achievement. Rock-climbing enthusiasts have told me that the Angel's 80-foot cliffs, towering sheer from the water, their ascent feasible only from the seaward side, have guaranteed them as much adventurous danger as anything they have ever undertaken. The reader will appreciate what this climb entails when one adds that this stack's circumference at the base is considerably less than it is halfway up! In other words, cliffs almost vertical, now begin to bulge—to overhang dangerously—as they reach up this alluring pinnacle.

With scientific precision, every avenue of the May's natural history has been explored. Scholarly *Transactions* have recorded its flowering plants and ferns, its fungi, its mosses, liverworts, and lichens. Likewise dealt with are its butterflies, moths, and dragonflies. My friend, W. S. Bristowe, has dealt with its spiders, not overlooking the harvestmen—the daddy-long-legs with which most of us have been familiar since childhood. Even its *ants* have received appropriate recognition.

Coming to the May's ornithology, one is overwhelmed by what its bird observatory, Britain's second oldest, has achieved. The range of migratory species recorded from the May since 1934 is astonishing. In the light of what already has been achieved in this field, it would be presumption on my part to trespass even upon its fringes.

In most Fifeshire homes situated along the twelve miles of coast between Elie and Crail, there linger traditionary fragments associated with the Isle of May. One such relates to the huge, blue stone by the Bluestane Hoose, just outside the Auld Kirk's gates, in Crail's Marketgate, where stood this historic community's first school. This stone, weighing maybe three tons, has been worn smooth by the generations of children that have clambered upon it, and perhaps also by equestrian ladies of an earlier era who used it as a mounting-block.

How came the Bluestane to lie where it does? Tradition answers that it was because of those unending squabbles between Auld Nick Himsel' and Crail's Auld Kirk minister. The conclusion was reached that only by banishing Auld Nick to the May was there any likelihood of restoring peace to the parish. While exiled there, he employed his time in aiming great, revengeful stones at the Kirk's steeple. All fell ridiculously short of their objective until, to the dismay of elders and parishioners, one dropped no distance from it. There it remains, exhibiting to this day the deep dent o' the De'il's thumb.

CHAPTER THREE

Orcadia's Perilous Waters

To THE north and north-east of that part of the Caithness coast spanning roughly 14 miles between Dunnet Head, in the west, and Duncansby Head, north-easternmost reach of the mainland of Great Britain, runs the Pictland, or Pentland, Firth, one of the most famous, most frequently traversed, and at the same time most dangerous of the world's major seaways. Immediately to the north, and at varying distances beyond its wild waters, lie the southernmost isles of Orkney. The Firth's width between Dunnet Head and Tor Ness, in Hoy, is just a little short of 8 miles. Between Duncansby Head and Brough Ness, in South Ronaldsay, it contracts to about 6. Its maximum width *between* these seaward extremities, however, is considerably greater, as at the Sound of Hoxa, beyond which expands the obsolete and almost entirely deserted naval anchorage of Scapa Flow, its waters so deep and capacious that, in the decades before air-power superseded sea-power, it would have accommodated all the world's navies.

Quite apart from the Pentland Firth's perilously powerful tide-races, its conflicting currents and turbulent roosts, running simultaneously in contrary directions at 12 or more knots, the navigator must eschew not only Stroma and Swona, two sub-stantial islands in their midst, but even more so the menacing cluster of rocks at its eastern approach known as the Pentland Skerries—the Muckle and the Little Skerry, Louther Skerry and Clettack Skerry. To guide him when harassed by what a chronicler of old referred to as 'ye sevene contrarious tydes of ye Pightland Frith,' Orkney's first lighthouse—that at North Ronaldsay—was established in 1789. The light on the Muckle Skerry came into service 5 years later.

With the Pentland Firth's most menacing areas, seamen of all the maritime nations must needs acquaint themselves. They must know precisely the whereabouts of the Boars, the combers curling ceaselessly between Duncansby Head and Stroma; the locality of tumultuous waters off St. John's Point, midway between Duncansby and Dunnet, known as the Merry Men of Mey; the great whirlpool off the north-west corner of Stroma, known as the Swelkie;[1] and also that off Swona, known as the Wells of Swona. Stroma's Swelkie is the most spectacular of Britain's whirlpools. The Corrievreckan of Hebridean waters is its only serious rival. It is a phenomenon of the *ebbing* tide just as the Duncansby Boars are of the *flowing*.

For good measure let us include also the Rispie's constricted tides as they charge irresistibly through the narrow channel separating the precipitous sandstone cliffs at the Knee of Caithness from the stack adjacent thereto. The height of the cliffs just here, and this stack's nearness to them, afford the most dramatic view of any tide-race in Britain. When this current is running swiftly northward between the Caithness mainland and this stack, its considerable downward slope immediately beyond the stack is plainly discernible from the proximate cliff-edge.

The perfect symmetry of the stack itself demonstrates the relative power of the sea and of the sub-aerial agents in the shaping of sea-cliffs. Whereas the waves and currents wear away the cliffs only at their base, the latter agents—air, wind, rain, springs, running water and frost—attack every vestige of the rest of them, particularly where the complete absence of vegetation leaves their rock surfaces entirely exposed to them. Where, as in the case of this stack, its sides recede uniformly toward the top, it shows clearly that the upper layers of the flagstone composing it have been worn away more rapidly by these disinte-

---

[1] 'A certain King Frodi,' according to one of the many Icelandic traditions recited in these northern latitudes, 'possessed a magical quern or hand-mill, called Grotti, which had been found in Denmark, and was the largest quern ever known. Grotti, which ground gold or peace for King Frodi, as he willed, was stolen by a sea-king called Mysing, who set it to grind white salt for his ships. Whether Mysing, like many another purloiner of magic working implements, had only learned the spell to set it going, and did not know how to stop it, is not stated. Anyhow, his ships became so full of salt that they sank, and Grotti with them. Hence the Swelkie. As the water falls through the eye of the quern, the sea roars, and the quern goes on grinding the salt, which gives its saltness to the ocean.'

grating agents than have been the lower layers by the marine erosion to which they unceasingly are subjected. Where the reverse is the case, cliff-tops overhang the sea. When one observes the force with which the waves and tides assault the base of cliffs as at the Rispie, one is astonished to find that these sub-aerial agents have shared in their erosion to so much greater an extent. Nowhere are the results of such erosion and disintegration more impressively—more dramatically—demonstrated than in the Old Red Sandstone sea-cliffs of Caithness and Orkney.

*   *   *

The Pentland Firth, after the Strait of Dover, is Britain's busiest seaway. Scan it for but a few seconds from John o' Groat's any day of the year, and a dozen or more vessels, ranging from small fishing boats to gigantic cargo ships and passenger liners, may be seen sailing eastward and westward through it. In addition to the considerable volume of British shipping based on the east-coast ports of the Humber, of Teesside, Wearside, and Tyneside, of Leith, Grangemouth, Dundee, and Aberdeen, all sea traffic between the Low Countries, Germany and Scandinavia, and the U.S.S.R., bound for or returning from North America, adopts this notoriously turbulent route—'everything from ocean greyhounds like the Swedish liners, *Kungsholm* and *Glipsholm* (over 30,000 tons) and Texan supertankers (60,000 tons or more) to 2,000-3,000-ton transatlantic tramps is seen,' as Bill Mowat puts it in his recent publication on John o'Groat's.

Having remarked on this mighty transverse shipping, it would be ungenerous to omit mention of the splendid services of the motor-vessel, *St. Ola*, Orkney's own very special mailboat which, undaunted by the fiercest tempests, crosses the Pentland Firth from Caithness's little port of Scrabster to Stromness, in Orkney's West Mainland, sailing early in the afternoon of every day of the year, reaching Stromness roughly two-and-a-half hours later and returning to Scotland the following morning. Those who physically run a service such as the *St. Ola*'s are but the merest fraction of the unsung fraction of mankind.

*   *   *

You will have noticed my reference to the *St. Ola*'s returning

to Scotland, as though Scotland were a different, if not actually a
foreign country—as in fact it is considered by most Orcadians.
Although five centuries have come and gone since King Christian
of Norway and Denmark ceded Orkney and Shetland to Scotland as part of the dowry of his daughter, Margaret, Orcadians to this day regard Scotland as a kingdom quite distinct
from their own, separated from their own by more than the
Pentland Firth. The extent to which this differentiation of theirs
persists can be seen in the way in which *The Orcadian*, Orkney's
weekly newspaper, deals with happenings on the farther side of
that famous seaway. For instance, one reads in that paper about
somebody—perhaps, a native on a brief visit to his relations—
who returned to Scotland by air or by the *St. Ola* on such
and such a day. News about Scotland—and I write this in no
spirit of criticism—is treated in its columns as of secondary
importance. It is news no less foreign than was that of the fall
of Sebastopol as it appeared in November, 1854, in the very
first issue of *The Orcadian*.

I remember a conversation I inaugurated with a quiet, sea-
gazing, Orcadian at Lyness, on the island of Hoy, by asking what
was the land we were seeing on the south horizon, several
miles away. I had uttered no more than the first word of my
proposed question when, having correctly anticipated it, he
immediately pointed south across the sea—south across the *Pictland* Firth, as some modern historians and cartographers
now designate it—and replied, as if anxious to inform me precisely of my whereabouts, 'That's *Scotland* you're seeing over
there!'

His remark reminded one of the establishment, about the middle
of the 18th century, of a regular ferry service rendered by six-
oared boats operating between Huna, in Caithness, and Burwick,
in South Ronaldsay. Although this service was superseded in 1856
by steamer communication, there were times when ferrying by
these oared boats had to be resorted to. A traveller to Orkney in
1860 recorded how, when bad weather prevented the steamer's
crossing, 'a ferry-boat carries the mails between South Ronaldsha
and Scotland'.

The Orcadian celebrity, Eric Linklater, expresses so much more
artistically what I am endeavouring to say:

'In Kirkwall there flourish musical and dramatic societies that offer imaginative entertainment on a high level of achievement; and a weekly newspaper, *The Orcadian,* is edited with a modern verve and skill that very cleverly present the matter of Orkney—its daily life and work—as necessary reading. *The Orcadian* pays scant attention to news from London or foreign parts, but never fails to report, in a very well-informed column, the seasonal activities of the innumerable birds that frequent its cliffs, and the arrival or departure of some rare migrant. A visiting film actress would not escape attention, but a hoopoe or honey-buzzard, a spoonbill or a little stint, would be more assured of a respectful paragraph'.[1]

* * *

Thousands of boats of one kind or another were lost in varying circumstances in Orcadian waters during the pre-historic millennia of primitive propulsion by oar, and perhaps by sail also. Several hundred are recorded as having been wrecked, either temporarily or totally, in historic times. Ships of the luckless Armada are known to have come to grief in these northern seas. In 1588 disaster befell the flagship, *El Gran Grifon,* commanded by Juan Gomez de Medina. In September of that year she ran aground in the Geo Swartz, a narrow inlet at the south-east of Fair Isle. Of her complement of 300 sailors and soldiers, roughly two-thirds were believed to have survived; and thus, according to an enduring tradition, the natives of Fair Isle obtained from shipwrecked Spaniards the patterns which have contributed so appreciably to the fame and ubiquity of their woollen goods.

A shipwreck of no less historical significance was that which occurred in December, 1679, six months after the Royalists' victory over the Covenanters by the banks of the Clyde at Bothwell Brig, where at least 1,200 of the latter were taken prisoner. In November of that year 300 of those refusing to submit to their conquerors' terms were shipped from Leith aboard the *Crown* for transportation either to the Barbadoes or to Virginia—nobody quite knows to which. What one *does* know is that there were only about 50 survivors when, the following month, this vessel was

[1] *Orkney and Shetland: An Historical, Geographical, Social and Scenic Survey,* published by Robert Hale in 1965.

wrecked at Scarva Taing, near Mull Head, a formidable promontory of Deerness, easternmost of the thirteen parishes comprising Pomona's, or Mainland's, 202 square miles.

Above the shore at Scarva Taing stands a monument commemorating the Covenanters' dire misfortune in waters nearby.
Accounts of what happened vary considerably. Robert Wodrow,
'the faithful and glorious author of the *History of the Sufferings of
the Church of Scotland*, who was born the year of the *Crown's*
disaster, gives a long and harrowing one. The *Diary* of Thomas
Brown, a Kirkwall lawyer, provides us with the most succinct:

'Ye 10th of Decr. 79, being Wedinsday, at 9 in ye evening or
yrabout, the vessell or ship callit ye Croun, qrin was 250 or
yrby of ye Quhiggs takin at Bothwall Brigs to have bein sent to
Verginy, paroched at or neirby ye Moull Head of Deirnes.'

*   *   *

As a result of two of those mighty wars in which Christendom
has specialised in our own century, Orcadian waters were the
scene of several major shipwrecks and much consequent loss of
life when the Germans managed to inflict upon our navy what we,
as proudly and zealously, would fain have inflicted upon theirs.
Evidence of this may be seen throughout Orkney. In the naval
cemetery at Lyness, on the island of Hoy, for instance, row upon
row of standard gravestones commemorates seamen of one kind
or another who perished there or thereabouts in war-time—boy
telegraphists, trimmers, firemen and stokers, engineers and engine-
men, petty officers, gunners, ships' stewards and cooks, ships'
carpenters, painters, greasers, and so on. Many such lost their
lives in the sinking of H.M.S. *Ajax* in April, 1915. Nowhere less
mistakably than in Orkney's burial-places do graven stones testify
to the unrepentant betrayal of what the Christians profess.
Scarcely an islet in these northern seas is without its memorialised
evidence of murderous wickedness blatantly perpetrated to the
Glory of God and of His Son, Jesus Christ, Our Lord!

*   *   *

One of the worst naval disasters in Orcadian waters during the
First World War occurred in heavy seas a mile or two off Mar-

wick Head, in the Birsay district of Mainland. There, at night-
fall on June, 5th, 1916, the cruiser, *Hampshire,* sank in twenty
minutes, with all but half-a-dozen of the 700 officers and men
aboard her, having struck a mine laid a week previously by the
German submarine, U75. Among those who perished were Field-
Marshal Earl Kitchener and his staff. A few hours earlier
Kitchener, then Secretary of State for War, had left Scapa Flow
where, aboard the *Iron Duke,* he had lunched with Admiral
Jellicoe. He was on his way to Russia to confer with the Czar.
Near Marwick Head, in the form of a small, rectangular tower,
stands the Kitchener memorial erected by the people of Orkney.
Inscribed upon a granite tablet inserted at its north side is the
sanctimonious twaddle in which the warring European nations
have specialised since Classical times.

Kitchener, since his death, has been as much a legendary
figure as the loss of the *Hampshire* has remained an unsatisfactory
mystery. Let us see what Eric Linklater has to say on this:

'In Orkney it was strenuously maintained that there might
have been many more if local people, who knew the shore, had
been allowed to help in the search for survivors. But some
perverted regard for "security" forbade all local assistance,
and some of the naval rescue teams were late in arriving because
they had lost their way or been misdirected.

Many foolish tales were propagated—there were those who
said that Kitchener had been deliberately sent to his death,
there were others who believed that he was still alive—and
the curious reticence of the Admiralty certainly deepened what
appeared to be a mystery. The few survivors who had reached
the cliffs, a little way south of Marwick, were not allowed to
be interrogated by the press, and it was commonly believed that
they had been posted, severally and apart, to remote stations
to prevent interrogation. A further difficulty was created by the
official report of the sinking; which, if my memory is correct,
was not published till long after the disaster. It was stated there
that the gale which was blowing when the *Hampshire* sailed was
north-easterly; but in Orkney it was confidently said to have
been north-westerly, and this seems to be substantiated by
the fact that the escorting destroyers had to be sent home

because of heavy seas. In a north-easterly gale the ships would
have been under the lee of the land from the south end of Hoy
to the Brough of Birsay. It is, moreover, hard to believe that
any survivors would have reached the cliffs if a violent, offshore wind had been blowing. Admittedly the tides in that
vicinity are strange in their behaviour, but I remember seeing
a Carley float—a kind of raft—jammed in the rocks a little way
south of the Head; and could the tide have carried so light and
buoyant a craft against a gale-force wind?"[1]

*   *   *

Among the several naval disasters which were to follow in these
waters was that which befell H.M.S. *Vanguard*. On July, 9th, 1917,
while at Anchor in Scapa Flow, she blew up, with a loss of almost
all her officers and men. Much less serious, of course, was the fate,
often without loss of life, of some of the small fishing-boats built
for or requisitioned for war service. There comes to mind the
Monarch, a herring-drifter built and donated by the women of
Chicago toward the end of the First World War. For more than
half a century she has stood conspicuously high and dry on the
beach at Herston, South Ronaldsay, no distance from the Hoxa
of Viking times.

Two years later Scapa Flow was to become the greatest ships'
graveyard in history. On June, 21st, 1919, the surrendered German
fleet, while anchored there, was scuttled. Seventy-four ships,
including eleven battleships, their seacocks deliberately jammed
open, went to the bottom. There they lay until 1924, the year
Messrs. Cox & Danks began their mighty salvage operations.
One by one these sunken ships were brought to the surface and
towed long distances to the shipbreakers' yards. Improvements
in salvage technique as the years went by made possible the
re-floating of such naval monsters as the *Moltke*, *Kaiser*, and
Hindenburg. I recall a visit I paid about this time to friends at
Inverkeithing. Then being broken up there, and at no distance
from their home, was a salvaged German destroyer, the name of
which I forget. On its encrusted deck, however, I walked leisurely
for a few minutes, reflecting upon the satisfaction man derives
from his zealous application to the potential destruction of his

[1] *Orkney and Shetland* by Eric Linklater (1965).

own kind. Man prefers revelling in his paradoxes to examining where they inevitably lead him. In the summer of 1939, while the last of the scuttled German vessels—the *Derflinger*—was being raised at Scapa Flow, a British ship, in preparation for yet another war, was engaged a mile or two away in laying a new defence boom across one of the channels giving access to it. Notwithstanding, on October, 13th, of that year, at high tide and midnight, a U-boat, slipping into the Flow between Lamb Holm and Glims Holm, torpedoed the battleship, *Royal Oak*, as she rode at anchor off Gaitnip, in Scapa Bay. A buoy marks the spot where she went down with at least 800 men. A few days later, the Luftwaffe circled over much of Orkney, so straddling with bombs the *Iron Duke*, Jellicoe's flagship, that she had to be beached. Otherwise she would have gone to the bottom.

To the several memorials already in Orkney's Cathedral of St. Magnus, there was added to its north aisle in 1948 a plaque commemorating the loss of the *Royal Oak*. Confidence in the security Scapa Flow hitherto had afforded the Home Fleet when at anchor there waned after this shattering blow to naval prestige, although, in order to reduce the likelihood of another such, several old ships were sunk in each of the sounds where they entered it from the east. These old ships still rust there, gradually sinking out of sight as the years go by. Subsequently, the four causeways known as the Churchill Barriers were constructed, for the most part of 5-ton and 10-ton blocks of concrete. The labour required for these causeways was supplied mainly by those Italian prisoners of war who erected on Lamb Holm, in this ultra-Protestant land, their artistic Roman Catholic Italian Chapel, now an Orcadian show-piece. Linking South Ronaldsay, Burray, Glims Holm, Lamb Holm, and Mainland, they were formally opened in May, 1945. They span channels with a maximum depth of over 50 feet. The longest measures nearly half a mile. Today they provide an uninterrupted motorway from the south of South Ronaldsay to the farthest Mainland homestead accessible by road.

In 1957 the Admiralty recognised as obsolete the Flow whence, 41 years earlier, the fleet under Admiral Jellicoe sailed to engage the German fleet which it did at the unrewarding Battle of Jutland—whence also set out so much that went into our equally disastrous Norwegian campaign. That year the naval base of

Lyness was closed down. This former destroyer base is now as much a ghost-town as is any of the Australian gold-rush towns I visited a few years ago. In the light of warfare as now envisaged, Orcadian waters are of little strategical value. To all intents and purposes, man meantime is done with them. They serve no longer his dubious requirements. The former scene of fabulous battleships and cruisers, of destroyers and minesweepers, is today the lonely haunt of our oceanic fowl. To me, who went to The Trenches in his teens 'to make the world safe for democracy,' it is immeasurably deplorable that we and the Germans, the two most intelligent and enlightened peoples on the earth, should have bashed one another to bits, and in doing so fractured irreparably the best of our waning Western Civilization.

*   *   *

So much for naval matters. Let me conclude with a paragraph or two on a couple of cargo-vessels that came to grief in Orcadian waters, the *Johanna Thorden* in January, 1937, the *Irene* as recently as March, 1969.

The former, a Finnish motor-vessel of 3,200 tons, was on her return maiden voyage from New York, when, on the murky night of January, 9th, and with a dangerous list to port, she came to disaster in the Pentland Firth. In two of her lifeboats everybody aboard her got clear. The first to quit, carrying 25 men, women, and children, was lost in the storm then raging. Three lifeboats (one from Wick, another from Thurso, the third from Longhope) searched in vain for her. The following morning she came ashore at Deerness, right side up, empty, and undamaged. The bodies of several of those she carried were afterwards picked up at varying intervals here and there along adjacent shores. The second lifeboat, with the captain and a dozen seamen, left the stricken ship a few minutes before she broke her back and sank. Only 8 of them reached shore alive. According to the account published in *The Orcadian*, the remaining 5, including the captain, were killed or stunned or drowned when their lifeboat perished amid the breakers off South Ronaldsay. One of the ship's survivors was her radio operator, who, for several years after his deliverance, used to holiday on Swona.

That murky night, as the natives of South Ronaldsay still recall, a woman well known to them had had a premonition about which she immediately told her neighbours. She had dreamt that the corpse of a fair-haired woman wearing gold earrings lay on the rocks by the shore, no distance from her cottage. Morning brought news of the *Johanna Thorden's* disaster and of the discovery on the rocks of South Ronaldsay of a drowned woman wearing gold earrings.

Exactly where, in the first instance, the *Johanna Thorden* struck, none of the 8 survivors of the total of 38 aboard her was able to say. Some thought she had hit one of the Pentland Skerries. Anyhow, she went ashore close to where she now lies submerged. That is to say, at a point roughly a hundred yards off the Tarf Tails, at the south-western tip of Swona, in that notorious tidal turbulence known as the Westerbirth. The information supplied to Lloyd's by the official receiver of wrecks as to where the sunken vessel lay was so vague that, but for the knowledge of one or two of Swona's inhabitants, all trace of her might have been lost. Lloyd's agent had placed her on the wrong side of the island. But her position to within a good stone's-throw has always been known to Rossie, a native who so readily relates how, just before she broke up to slither back into the sea, he rowed over to her and retrieved the packets of cornflakes he so greatly enjoyed.

Encouraged by improvements made in diving technique during recent years, and also by the greater ease with which wider areas of the sea's bottom can be searched, Keith Jessop and Frank Guest, of Keighley, established in 1968 their Western Isles Diving Services with a view to salvaging commercially from wrecks of this kind what they might yield. They knew that the *Johanna Thorden's* cargo included a considerable quantity of copper in coils, bars, and billets, all of which in 1937 had been written off as a total loss. In March, 1969, and in a cold offshore north-westerly—the most suitable wind for underwater operations in this sea area—they located this wreck after much strenuous searching. With her underwriters they promptly came to terms, resolved to recover what they could of the copper that went down with her. At the same time they agreed terms with that well-known Stromness character, Ginger Brown, in regard to their engaging his services with his motor fishing-vessel, *The Three Boys*. Ginger, in addition to his owning this 2- 3-tonner, so well adapted to an enterprise of this nature, is an asset to them because of his once having been employed by Cox & Danks. Working meantime at a depth of 60 feet on the copper lying in what would have been the ship's forward hold, they recover, on days when diving conditions are favourable, an average of 25 hundredweight. Owing to the steep angle at which the wreck now lies, the rear part of her copper rests at 70 to 90 feet.

Off Swona, past which the spring tides sweep at 10 to 11 knots, salvage operations are feasible only at the flood—only when the tide is flowing from east to west through the Pentland Firth. All such matters the divers study meticulously, seizing every opportunity of descending to the wreck with their tackle. To a large, securely padlocked shed at Scapa pier, *The Three Boys* carries each day what they have raised. They tell me that, when they have succeeded in removing from this wreck all the copper they can, they mean to turn their attention to *other* wrecks lying out of sight in Swona's neighbourhood.

*   *   *

The worst misfortune Orkney has suffered in recent years is that which, in March, 1969, ensued from the difficulty into which the 2,000-ton Liberian-registered and Greek-owned *Irene*, well-found and in good shape after a costly refit, ran when she got out of control in Pentland waters and drifted helplessly before the south-westerly gale that had been raging for several days, whipping up tremendous seas rendering 10 days overdue the routine relief at the Pentland Skerries' lighthouse. The *Irene* in ballast, and with a crew of 17 Greeks, Algerians, and Portuguese, was on a voyage from Granton to Christiansund. Before she finally went on the rocks near Grim Ness, where, broadside on and on an even keel, she lies hard and fast, and is likely to remain, she had radioed her distress. In response to this and to the flares she fired, lifeboats from a number of stations put out to her, among them that from Longhope which, in the course of her rescue mission, capsized and sank with all hands. A wave of enormous dimensions is thought to have been responsible for this truly shocking fatality. Of her crew of eight, two were fathers, and four were those fathers' sons.

After a simple service in South Walls parish church, the bodies of seven of them (that of the eighth had not been recovered) were laid in Osmundwall cemetery, situated a little above the historic inlet of Kirk Hope, the Asmandarvagr of the Sagas. There, nearly a thousand years ago, the Viking, Olaf Tryggvason, King of Norway, offered Earl Sigurd of Orkney and his followers the alternative of Christianity or death by the sword. The lifeboatmen's tragic end widowed seven and orphaned ten. It brought messages of sympathy from all over the world, even from remote Tristan da Cunha.

Longhope's lost lifeboat's record in recent decades is, indeed, an impressive one. In April, 1951, she rescued 13 men from the stricken trawler, *St. Clair*. When, later, the enormous Swedish tanker, *Oljaren*, ran aground on the Skerries, she took off her 40 men in an operation lasting 32 hours and necessitating three trips to her. Later again, when the Norwegian ship, *Dovrefjell*, went on the Skerries, she rescued 41 of her crew. One of her most celebrated achievements was her rescue of the crew of the trawler, *Strathcoe*, fast under the mighty sea-cliffs of Hoy. Early in 1962 this lifeboat proceeded far out into the North Sea to rescue the crew of a small steamer called the *Daisy*. The crew, however, had taken to the life-raft picked up by another ship and transferred later to the Longhope lifeboat. More recent were her services to the stranded *Ross Puma* and the *Ben Barvas*.

The entire complement of the *Irene*, whose call for help Longhope's mighty men of valour had answered, was brought ashore by breeches buoy operated by the life-saving teams of St. Margaret's Hope and Deerness. Indeed, these rescuers were already waiting for the *Irene* on the shore where eventually she grounded.

Not within living memory has there gathered in Orkney's own Cathedral of St. Magnus so large a congregation as attended the Longhope lifeboatmen's memorial service held there the following Sunday morning. It was estimated at well over a thousand.

Eight centuries earlier, this mellowing and remarkably beautiful edifice had been built by men familiar with these perilous waters.

CHAPTER FOUR

Shetland:
Isles of the 'Simmer Dim'

How SIMPLE today, if not also surprisingly abrupt, is the speediest method of reaching Scotland's Viking Isles, those northern outposts ceded by Norway in 1569, when James III of Scotland married King Christian I's daughter Margaret. Less than an hour's exhilarating flight from Aberdeen under B.E.A. auspices, and but thirty minutes' from Kirkwall, in Orkney, carries one over that turbulent tide-race, the Sumburgh Roost, to the runway just beyond Sumburgh Head and its Stevenson lighthouse. This is the southernmost reach of Mainland, largest by far of the hundred islands and islets comprising Zetland, or Shetland, most northerly of Scotland's 33 counties and, together with Orkney, northernmost of her 71 parliamentary constituencies. These islands' total area is 550 square miles, or 352,000 acres. Less than a fifth of them share their population of roughly 21,500, six-sevenths of which occupy Mainland's 265,000 acres—a low average of inhabitants to the acre.

Travellers with a preference for voyaging in the old, romantic style will embark any week-day for Lerwick, Shetland's capital, either at Leith, or at Aberdeen, whence runs one or other of the North of Scotland, Orkney, & Shetland Shipping Company's twin-screw motor-vessels, *St. Clair* or *St. Ninian*. Sailing from the latter, one reaches Lerwick in about 14 hours, having covered a distance of 180 nautical miles—only 24 miles fewer than the distance across the North Sea from Shetland's Out Skerries to the Norwegian coast.

From Sumburgh airport, Mainland, elongated and greatly indented by ever penetrating voes, stretches north a distance of

ISLE OF MAY. The Tower, the Isle's present-day lighthouse

ISLE OF MAY. The Beacon – all that remains of Scotland's first lighthouse

ISLE OF MAY. The gullery on Rona

ORKNEY. The Kitchener Memorial at Marwick Head, commemorating the loss of the *Hampshire*

ORKNEY. The wrecks of vessels deliberately sunk in one of the seaways formerly giving access to Scapa Flow

60 miles to the rocky extremity of North Roe. Reaching farther
north still are Yell and Unst. The former island, bleak and
barren, measures 17 miles from south to north. With an area of
roughly 52,000 acres, it is the second largest of the Shetlands.
The latter, with a maximum length of 12½ miles in the same
airt and an acreage of 30,000, comes next in order of size.
Muckle Flugga, a skerry lying a thousand yards off Hermaness,
the northern part of Unst and now a bird sanctuary, is the second
northernmost speck of the British Isles. On it stands the remote
North Unst lighthouse. At a distance of a little over 700 yards to
the north-east of it rises a black rock, usually awash, shaped like
a whale's back. This is Out Stack, northernmost fragment of
Britain. Beyond lie nothing but water, the Arctic ice, and the
North Pole.

Few realise that almost the whole of the Shetland Isles' south-
north length of approximately 70 miles lies north of the 60th
Parallel. Only its southernmost 10 miles reach farther south
than does Cape Farewell, the most southerly point of Greenland.
Nevertheless, owing to the Gulf Stream's North Atlantic Drift,
seldom is the climate of the Shetlands truly cold, even in mid-
winter. January temperatures at Baltasound, on Unst, most
northerly of the Shetlands, are comparable with those at Kew. On
the other hand these islands are often swept, and for considerable
periods, by gales of tremendous force. 'In January or March,' as
Eric Linklater put it, 'a benighted traveller may be felled by the
wind, but he will not be frozen.'

After Unst in order of size comes Fetlar with its 10,000 comparatively fertile acres, famed for their Shetland ponies. Two
somewhat smaller islands might be mentioned, namely, Whalsay,
lying some 14 miles north-east of Lerwick, and Bressay, but a
mile to the east of it, the narrow sound intervening between them
affording at all times safe harbour and convenient anchorage for
the sadly declining fishing fleet operating from Lerwick, as well
as for the sundry ships of many a maritime nation calling during
this port's busiest months, or anchoring in sheltered Bressay Sound
in time of storm.

Bressay, with an area of 7,400 acres, carries a population of
nearly 6,000—roughly the same as Lerwick's at the height of the
herring-fishing season. Its fishing history goes back to the middle

of the 15th century, when Dutch luggers assembling in its sound
during the month of June, at the beginning of the season, made
Bressay the headquarters of their activities, and were in large
measure responsible for many of its oldest, stone-built, waterfront
buildings—buildings that to this day, like those deceivers about
whom Shakespeare wrote, have 'one foot in sea, and one on
shore'. During the 16th and 17th centuries as many as 1,500
of these luggers might have been counted in this sheltered and
comparatively narrow seaway, always the scene of ardent activity,
and often of international rivalry. In 1652, when the Common-
wealth was at war with Holland, 94 English ships under Deane
and Monk anchored in Bressay Sound. The following year a
fort was erected on this channel to control it. In 1665 another
English fleet, one of 92 sail under Lord Sandwick, reached these
waters to drive away the Dutch. Roughly a century later—in
1753, to be precise—the frigate, *Wolfswinket*, engaged in pro-
tecting 500 Dutch herring-boats in these waters, was attacked
by 4 French ships. Rather than surrender after several hours'
bloodshed, de Bardt, the frigate's defiant commander, blew her up.

Life on these islands has never been free of peril and adversity. Such lives and resources as commercial rivalry had not
destroyed, violent storms often did. One summer's night in 1881,
while a large fleet of sixerns from Northmaven, Yell, and Unst was
fishing some 40 miles north of the last mentioned, a sudden
storm necessitated a run for shelter. Ten boats and 58 lives
were lost, most of them off the north of Yell, between the Holm
of Gloup and the adjacent voe, where they might have found
survival.

A more recent mishap at sea which comes to mind is that
which, in 1930, befell the *St. Sunniva,* a vessel that at the turn
of the century had taken a prominent part in the endeavour to
establish with the Shetlands a Norwegian fiord passenger traffic.
In a dense fog she struck a rock off Mousa and went down,
but without any loss of life—most unusual when anything goes
wrong in these perilous seas.

Since even the briefest résumé of Shetland's recorded shipping
disasters would require several pages, I must limit myself meantime
to mentioning just one more. At the seaward end of the little
peninsula of Scatness, not far from Sumburgh's links and airport,

stands the old home of Betty Mouat. In 1886 Betty, then an
invalid of 60, set out for Lerwick aboard the sailing-smack,
Columbine. When its skipper was accidentally knocked overboard, the crew of two tried with the small boat to retrieve
him. All three were lost. Wind and tide bore the smack across
the North Sea while for eight stormy days steamers searched for
it. Eventually it was cast ashore on the Norwegian island of
Lepsoe, from which a rope thrown to Betty enabled her to land.
Shortly afterwards she was shipped back to Lerwick, on her native
isle, having by this time recovered from her ordeal.

Penetrated so deeply and numerously by voes—by creeks and
inlets—are Shetland's Old Red Sandstone cliffs that nowhere
inland is the sea distant more than two or three miles. This means
that, in relation to area, no land surface in the world has a
lengthier coast-line, a coast-line increasing, of course, with
every inward encroachment by the sea. It also means that nowhere
can one proceed far enough from the sea to be able to refer to
one's position as inland.

No coastal stretch in Britain exhibits more recent, more
extensive, and more dramatic evidence of the violence with
which raging tempests from all directions have assailed it. Particularly is this so on the islands' west side, where they have
produced scenes of desolation indescribable, having torn out
enormous rock masses and transported them, a few to as much
as 100 feet, before shattering and weathering them into the
weirdest shapes.

That for centuries the Shetlanders have taken advantage of the
sea's proximity to their hearths is seen in the fervour with which
they have prosecuted the fishings. Until about fifty years ago,
the herring trade brought to them an uninterrupted and fairly
well distributed prosperity. The herring industry's concentration
on Lerwick, rendered inevitable by the sudden and extensive
introduction of the steam drifter, altered things. Nowhere is there
clearer evidence of this than at Balta and Baltasound, situated off
the east of Unst. There, to this day, stands the ruined parish
kirk formerly packed to overflowing on summer Sundays by the
fisherfolk employed at no fewer than forty-five curing stations,
now derelict among the sound's fringes. During the summer
herring season in the years preceding the First World War,

fishing-boats in their hundreds frequented Baltasound where, often without intermission, curers worked consecutive rounds of the clock.

Dutch, Norwegian, Danish, and Swedish boats for centuries have frequented Shetland's fishing-grounds. They and their crews are still a common feature of Lerwick at the height of the fishing season. Many a sturdy, wooden, steam- or oil-driven craft from their enterprising lands is to be seen there, year after year. Just as the natives of Barra recall the days when the wherries and herring drifters at anchor over the weekends filled Castlebay's splendid harbour, and appeared to reach as far as Vatersay, the Shetlanders speak of a time when *their* herring fleet extended all the way from Lerwick to Bressay. Today, along Bressay's inward shore, as likewise along the fringe of Castlebay's harbour, one sees the decayed piers and stages at which herrings were landed and cured in those bountiful times.

No less prosperous then was Scalloway, once the capital of the Shetland Isles. But Lerwick's rapid commercial development wrested from Scalloway this distinction. Herrings are still landed there, of course. Apart from such of them as are kippered locally, they are now transported by motor-lorry seven miles over the intervening high ground to Lerwick. Scalloway remains the centre of Shetland's white fish industry, however.

Throughout the Second World War, Scalloway was a secret Norwegian base. During the German occupation of Norway several raids upon its coast were made from Scalloway by the Norwegian force stationed there. In escaping from the Nazis, these refugees found their way across the North Sea in their small ships, as did their forefathers when fleeing a thousand years earlier from the oppressive Harald Fairhair. Norwegian crews engaged in shark-fishing in these northern waters in pre-war years were, of course, already acquaint with Scalloway's harbour.

Boats from Norway still visit this little port, contributing to its boom and bustle. Several of the Norwegians stationed there married Shetland lassies, with the result that boats from 'Norraway ower the faem' often have a personal reason for calling there. Today, incidentally, this industrious locality of some 600 inhabitants is noted for its boat-building and boat-repairing. The keel of many of the smaller craft sailing our northern seas was laid at Scalloway.

A fine, old house of stone, with windows remarkably small for its size, soon arrests the attention of anyone exploring this village. It is known as the Muckle Haa, the Great Hall, and is now let off in apartments. All its rooms are beautifully panelled. Like the rest of the solid, stone homes and buildings in these Isles, it was built to withstand the worst that the North Atlantic's storms can deliver.

The most prominent and celebrated of Scalloway's ancient monuments is, of course, its ruined castle of corbelled turrets. It was erected in 1600 at the head of its peninsula by the despotic Patrick Stewart, Earl of Orkney, to replace his earlier residence at Sumburgh, the foundations of which, resting precariously upon sand, began to subside. Patrick, known to local history as Black Pate, was the son of Robert Stewart, Earl of Orkney, an illegitimate son of James V of Scotland. In 1591 he succeeded his equally tyrannical father to wield an uncompromising authority alike over his Orcadian and Shetland subjects, driving them to quarry and to transport stone and lime for his stronghold —compelling them, as history relates, to undergo every manner of painful servitude, and 'without either meat, drink, hire, or recompense of any kind'.

The life of the Shetlander is sufficiently different from that of the Orcadian to warrant the popular description of the former as a fisherman with a croft, and of the latter as a farmer with a boat. Such of the Shetlands as are still inhabited have a total of over 2,000 crofts. No less apt a differentiation is the modern one —Orkney for ancient monuments: Shetland for cliff scenery. The cliffs one immediately recalls are those attaining at Fitful Head a height of nearly 1,000 feet. Equally spectacular, although roughly 400 feet less lofty, are those at the Noup of Noss, that scene of incredible prodigality when myriads of seabirds are occupying in noisy clamour their traditional nesting-sites overhanging the surging sea.

Whereas in winter-time the Shetlands' northern latitude allows but five hours' weak daylight, it also ensures for them their nightless summer—the 'simmer dim', the twilight interval between sunset and sunrise, lasting for about an hour and continuing for a

month or so. During that time of the year there is no darkness
whatsoever. Anybody familiar with these northern latitudes in
midsummer is enchanted by this afterglow, this gloaming, just
as of old were the Shetland folks who, during this witching hour,
when the trows came out to revel in 'da dim', kindled the bonfires
round which both they and the trows danced and made merry.

Shetland's 400 miles of quite commendable roads are largely
coastal and elevated. Where not so, they traverse moorland territory grazed by the tens of thousands of sheep which have given to
these islands their wool and their famed hosiery—territory
intricately diversified by hundreds of peaty pools, and by lochs and
lochans and streams so stocked with trout as to have made Zetland
the anglers' paradise.

Shetland certainly is not as devoid of ancient monuments as
the saying quoted earlier would lead the uninformed to suppose.
On the contrary, its finest examples of these, pleasantly accessible
and now internationally recognised, are of considerable significance since they cover clearly all the archaeological periods from
the Neolithic (about 2000 B.C. in Scotland) to the Middle Ages.

On what remains of a grassy mound situated roughly a mile
north of Sumburgh Head, overlooking the West Voe where the
cliffs decline to a shallow bay and a flat, sandy isthmus, are the
late 16th-century ruins of Old Sumburgh House—'da Laird's
Hoose', or Jarlshof, as it has been called since Sir Walter Scott
gave to it in *The Pirate* its earliest footing in literature. Knowledge
of its existence and an assessment of its value to archaeology date
from the closing years of the 19th-century when storms attacking
the southern part of the mound revealed in its eroded, seaward
face portions of massive stone walling. Between 1897 and 1905 the
late Robert Bruce, owner of Sumburgh, traced these to the base of
some mediaeval buildings, and in so doing brought to light part
of an Iron Age settlement which included a broch, wheelhouses,
and a number of passage dwellings. Not until 1925, when the site
came into the hands of the Commissioners of Works, were further
discoveries made. Systematic exploration between 1931 and 1935,
under the guidance of the late A. O. Curle, at one time director
of Scotland's National Museum of Antiquities, and subsequently
of the Royal Scottish Museum, disclosed a large segment of a
Late Bronze Age and Early Iron Age settlement. Exploratory

trenches dug in 1933 revealed the presence of the third settlement, which dates from Viking times. Subsequent investigations
by Professor V. G. Childe, Miss B. Laidler, and Dr. J. S.
Richardson, involving the stripping of the Norse farmsteads and
examination of deeper and older occupation levels, were interrupted by the outbreak of the Second World War. Between
1949 and 1952 the Ministry of Works carried out the final phase
of excavation. This elucidated the history of a unique com-
mixture of settlements inhabited over a period of more than
3,000 years, contributed materially to our knowledge of the Iron
Age village, and enabled J. R. C. Hamilton, Inspector of Ancient
Monuments, to present us with his meticulous report on one of
the most remarkable archaeological sites ever excavated in the
British Isles.[1]

Of the broch at Jarlshof, oldest of its ancient structures, only
two-thirds remain. Circular in plan and roofless, as are all brochs,
and with a wall 17 feet thick at the base, it has a diameter of
nearly 30 feet. When complete, it may have exceeded 40 feet in
height, as did the broch on Mousa, an island half a dozen miles
toward the north. The latter is the best known and by far the best
preserved of all Britain's brochs. Certainly very much more so than
the example at Clickimin, on the outskirts of Lerwick, it rightly
ranks as the most outstanding antiquity of its kind in existence.

Traditionally and popularly, the brochs have been attributed to
the Picts, who are said to have sought refuge in them at the
approach of Viking sea-rovers. These structures are frequently
referred to as Pictish towers, or forts, and are sometimes entered
on maps as such.

Once a year the Shetlanders, demonstrably mindful of their
Viking ancestry, rejoice and make merry in a truly spectacular
fashion when, in Lerwick, on the last Tuesday in January, they
celebrate in the Viking fire-festival of Up-Helly-Aa the end of
the winter solstice and the return of spring. Until 1889 this was
done with tar-barrels set alight. Today it is the splendid proces-
sional occasion when an elected Guizer Jarl, clad in full Viking
costume and complete with helmet, battle-axe, and shield, stands
at the steering-oar of the richly coloured model of a Norse war-

---

[1] *Excavations at Jarlshof, Shetland,* published in Edinburgh in 1956 as
the first of a new series of Ministry of Works Archaeological Reports.

galley some 30 feet long, decorated with Norse symbols—a dragon head and tail at stem and stern, raven banner a-fluttering at the mast-head, shields bearing heraldic emblems arranged along the gunwales, each of the dragon's fearsome eyes lit up from within by a powerful electric torch.

Toward evening the galley, ringed round by cheering and shouting guizers, is hauled to its burning-place by the sea's edge. There, to the singing of *The Norseman's Home*, it is set alight as each guizer, following the example of the Guizer Jarl, flings into it his flaming torch, commemorating the days when fire cere-moniously consumed the deceased Viking leader along with his favourite galley. Thereafter, well into the small hours, feasting and dancing go on in every hall throughout the town. According to the Official Guide, published with the authority of Zetland's County Council and Lerwick's Town Council, the costumes to be seen on this annual festival occasion rival those at the Chelsea Arts Ball.

Few regions so sparsely endowed by Nature with riches as commonly understood and coveted offer such a range of whole-some recreation as these isles of 'the simmer dim'. They are essentially a territory for the archaeologist and the antiquarian, for the historian and the folklorist, for the artist and the photo-grapher, for the geologist and the ornithologist, for the angler in pursuit of brown and sea trout, for the yachtsman seeking regatta status.

These Isles of the Nightless Summer and of the virtually Dayless Winter, a territory of some severity and austerity, are inhabited by a hardy, industrious, courageous, and well-read people. The bibliography drawn up a few years ago by G. W. Longmuir, Zetland's Librarian, discloses at a glance its inhabi-tants' penchant for scholarship and literature. Many of them received their education at Lerwick's admirable secondary school, the Anderson Institute, and at Aberdeen University. The former was the gift of Arthur Anderson, a native who rose from the humblest beginnings to found the P. & O. These descendants of the roving, rieving Vikings write English with a fluency, purity and beauty truly remarkable.

ORKNEY. The Liberian-registered and Greek-owned cargo vessel, *Irene*, fast on the rocks near Grim Ness, on the south side of South Ronaldsay since March 1969

ORKNEY. Looking across Kirk Hope, the Asmandarvagr of the Sagas, to Osmundwall cemetery

SHETLAND. Scalloway's most pro ancient monument is, of course, its castle of corbelled turrets. This graph was taken during 'the simm

SHETLAND. The Broch on the Mousa, 'the best-known and by best preserved of all our brochs'

SHETLAND. 'Mony o' the auld ho Lerwick hae ane foot in sea an shore'

CHAPTER FIVE

A Journey to Noss

EVERY CLOUD, they say, has a silver lining. Certainly, the cloud of disappointment that overshadowed me the day I sought to re-visit remote Mingulay, in the Western Isles, was lined with something compensating, with something more argent than I could have anticipated among the Northern Isles, my acquaintance with which, hitherto, had been but slight. This was in the month of June, 1949. In assessing the good fortune now to be my lot, I recall that this northern expedition arose entirely from the escapade of a stirk—'the wee stirkie that got away on them,' was Alick MacLean's mother explained by way of an excuse for Alick's dereliction of promise. Alick was under obligation to transport me from Castlebay, in the island of Barra, to Mingulay, in order that I might again review the myriad seabirds then nesting on its formidable cliffs. The boat in which we should have sailed together belonged to Peggy Greer, a friend of mine who, at the time, owned both Mingulay and Berneray. Peggy employed Alick MacLean as boatman and as supervisor of her sheep-stock on Mingulay. She had told him that, as soon as weather permitted, he must land me there as early in the day as possible, and leave me ashore for ten or twelve hours at least, if not actually overnight.

Suitable weather conditions followed almost immediately. Calm water, bright sunshine, and towering cumulus augured well, as Alick could have seen from his own doorstep, a few hundred yards away from my own at the time. But, as his mother explained when I arrived on her threshold with camera and provisions, Alick and his neighbours were off to the hills. A panic-stricken stirk, while being loaded on to a boat at Castlebay, had bolted to freedom. As a vegetarian, I could hardly have blamed it.

Realising that my chances of getting to Mingulay were now pretty small, I decided to waste among the unreliable and procrastinating Celts of the western seas no more of such time as the advent of exceptionally good photographic weather had made the more valuable. I would try, instead, my luck with the Scandinavian stock of the *northern* seas. I would make Noss my objective. If this weather held, the teeming bird-life at the Noup of Noss, justifying so amply the island's having been declared in 1955 a national nature reserve, would furnish me with photographic opportunities sufficiently gratifying to warrant an immediate change of plans. I had long contemplated such an excursion to the Shetlands, anyhow. The stirk's very natural behaviour now provided me with the pretext. I promptly wired from Castlebay to B.E.A. at Renfrew, explaining my plight and my requirements. I obtained as promptly not only a flight by that morning's plane from the smooth sands of the Great Cockle Shore of Barra, but also the promise of an immediate flight from Renfrew to Sumburgh, Shetland's airport. Confirmation of the latter awaited me when, less than two hours later, I landed at Renfrew, and was met there by the press relations officer, who knew the Shetlands well.

'You're booked for Sumburgh all right!' were the words with which he greeted me. 'According to weather reports just received, it's bright sunshine up there, though, as you see, it's pouring buckets here. So, don't be too depressed. You'll get photographs on Noss all right!'

Forthwith he conducted me to the Dakota that awaited me. In a moment I was airborne again, and completely re-orientated. 'We're halfway to Inverness, sir,' the steward announced when we had been but ten minutes aloft. 'We're over Loch Rannoch, just halfway.'

It would have sounded foolish to have enquired of him how, surrounded on every hand by dense cloud since leaving Renfrew airport, he knew so precisely our whereabouts. Sunlight seemed as far away above us as the earth lay beneath us. Nothing was visible except our own aerial vehicle and the thick, dank whiteness it penetrated so impersonally, so unemotionally.

At Dalcross, Inverness's airport, a locality I had known intimately in boyhood, dismal rain was falling as we touched down. Here we were joined by a minister, a native of Shetland flying home on a month's holiday. He sat himself down beside me and, throughout the twenty minutes' flight to Kirkwall, our next stop, bemoaned an invisibility precluding his spotting, for my edification, prominent landmarks with which he had made himself familiar through his not infrequent trips between his charge near Inverness and his parents' home in Shetland. Never before had he not seen John o' Groat's when flying over it: never before had he been unable to distinguish certain components of the Orkneys when approaching Kirkwall. Nevertheless, our wings were a dazzling silver as we flew in over Sumburgh Head, the most southerly headland of Shetland. By now we had left dark clouds behind us to fly north into the gleam of silver linings. Far away, when just beginning our descent to the airport occupying the flat expanse behind this headland, I picked out, ahead of us, my objective, the Isle of Noss, its sloping plateau, and also the Noup to which this plateau ascends, clearly discernible to anyone knowing precisely what he was looking for. Now within sight, at a distance of twenty-two miles, it might well be within reach of Lerwick on the morrow.

Below us now, sweeping round Sumburgh Head, was that 'scene of ocean turmoil' as a voyager once described it, that menacing current known as The Roost, scene of many a disaster in the days of sailing-ships and, incidentally, *locale* of the wreck from which Sir Walter Scott arranged that the gallant Mordaunt Mertoun should rescue the buccaneering Clement Cleveland. A few seconds later we touched down at Sumburgh airport, a legacy from the Second World War, its runways intersecting at all angles the levelled expanse bordered at its sea-fringes by sand-dunes and marram grass, a region renowned in pre-war years among botanists who arrived by steamer at Lerwick from every part of the world, and travelled thence by road to Sumburgh. From late in May, right through the summer and autumn months, many varieties of wild flowers still thrive hereabouts, in spite of the drome now occupying so much of their native domain.

Our having landed at Sumburgh some little time before the bus was due to leave for Lerwick afforded me an opportunity of re-visiting, though only for a few minutes, Jarlshof, already referred to, one of the most edifying archaeological sites excavated in the British Isles, situated by the shore at no distance from the airport. Sir Walter Scott, in *The Pirate*, wrote graphically of it as the home of the Mr. Mertoun alluded to earlier.

*   *   *

I was getting on nicely with my camera among these ancient foundations when the blast from an impatient motor-horn summoned me back to the airport, whence the bus was about to leave. My clerical fellow-passenger was now to seat himself beside me in the bus ready to convey us northward to Shetland's capital. This enabled him to announce in my ear the name of every farm and crofting township, of every voe and hillock, that came in sight as we travelled along. If bad weather hitherto had denied him the pleasure of displaying his encyclopaedic knowledge of our whereabouts while airborne, sunshine now provided him with this pleasure while bus-borne.

'You'll be going to the Queen's in Lerwick?' he ventured. It happened that I *was* going there.

'Mony o' the auld hooses in Lerwick, as you'll see for yoursel, hae ane foot in sea an ane on shore, as Shakespeare says. The sea comes right up tae the foondations o' the Queen's. If you're a fisher, you could easy catch fish frae your bedroom window if it's on the seaward side.' As Dr. Harry Lillie puts it in that splendid book of his, *The Path Through Penguin City*,[1] if you really must have fresh piltack—fresh coal-fish—for breakfast, you can catch it so easily from the windows of the Queen's that, provided your rod be long enough, you may do so without leaving your bed! One recalls in this context Pliny's description of Comedia, his villa situated so close to Lake Como that 'almost from one's bed one can fish as from a small boat'.

As we approached the crofting township of Sandwick, our cleric, wild with enthusiasm, bubbling over with information of which he must needs divest himself, drew my attention to the Isle of Bressay. 'You'll see it faur better frae Lerwick,' he continued. 'And you see yon high bit? Yon bit away tae the right? Well, that's Noss—a bird sanctuary. Yon highest bit is the Noup

[1] Published by Ernest Benn in 1955

—the Noup o' Noss. Man! you never saw anything like the seabirds yonder!'

I now had to confess that a sight of the Noup and of these very birds was the object of my excursion to these northern parts; but I did not have the heart to add that, magnificent as the scene must be, it couldn't compare in stature with scenes of the kind I already had witnessed on Mingulay and on St. Kilda.

'And you see yon island there?' he continued, pointing to an isle but a couple of miles to the east of us. 'That's . . .'

'Mousa!' I interposed.

'How did you know that, now?' he asked in an astonishment which increased when I added that I was familiar with Mousa, having once spent an entire day ashore there.

'You'll know all about its Pictish broch then,' he continued, as this famous remnant of antiquity came into view.

'Indeed, I do!' I replied. 'I've known about it ever since my schooldays at George Watson's. Hume Brown's history-book had quite a bittie on it, and also a wee photograph of it.'

Though today Mousa is uninhabited, it once carried a population sufficiently large to have warranted its having a mill of its own, and also a commodious kiln in which the islanders in olden times dried their corn. Mousa is now used solely for grazing, though evidence of pristine tillage is still to be seen on every hand. At its stone jetty small boats disembark archaeologists arriving from every part of the world to examine its celebrated broch, which has been preserved almost in a complete state, and is undoubtedly the best known antiquity of its kind in existence. Few monuments of its age enjoy the good fortune of having suffered so little at the destructive hands of man.

*　　*　　*

All this time, the Isle of Noss and its birds were coming increasingly within my reach. From the window of the room allocated to me at the Queen's, as I immediately perceived, one so easily could have cast a line into the sea below. From that window, moreover, I could see quite plainly, roughly four miles to the east, where Noss, at its seaward extremity, terminates in its towering Noup, whence it sends its bird-haunted cliffs perpendicularly to the Atlantic, 600 feet below. Although this drop is less than half that of the western cliffs of Mingulay, I knew that now, at midsummer, it would be spectacular and dramatic enough, its nooks and crannies and ledges of weathered sandstone carrying, in proportion to their area, a dense population of nesting seabirds—of guillemots and razorbills, shags and storm-petrels, puffins and gannets, and every species of gull.

The greater black-backed gull nests at the top of Noss's cliffs and on its stacks. The small, black guillemot, which is quite common here, nests in cracks in the lower cliff-tops. The puffin nests in various parts of the isle, as do also the Arctic and the common tern, the former with its blood-red beak, red right to the tip, the latter with an orange-coloured beak tipped in black. These terns otherwise are much the same in appearance. They nest in mixed colonies.

What would these Shetland Isles be like without their birds? What would Lerwick be without its herring gulls, screaming in fierce competition over edible refuse tipped overboard into the harbour from trawler or drifter, or going noisily through the town's garbage bins and pails before perching in silent, vigilant rows upon the roof-tops?

The prospect of my reaching the cliffs of Noss the following day was sufficiently rosy to have smoothed away the disappointment I had sustained that morning in regard to Mingulay, now so distant. For all I cared, Alick MacLean and his neighbours could go on looking for the wee stirkie as long as they liked. Shining wings had borne me swiftly and safely from their aggravating undependability. On consulting the Barthomolew Half-inch I had brought with me, I saw that from Lerwick I would have to cross Bressay Sound to the island of that name, and then tramp two miles due east across that island to Noss Sound, beyond which, at its narrowest, the isle of my quest lay only a couple of hundred yards. Whether I might find anyone who would ferry me over the latter sound, no one could say. The shepherd, Noss's solitary human inhabitant, might just do so, provided I could shout loudly enough from Bressay's eastern shore to attract his attention, and wind and tide were such as would make ferrying feasible without undue hazard.

Noss, as one recognises in its ruins and in the grass-grown foundations of former homesteads, once carried quite a population.

Among those who spent their youth there was Dr. Copeland, whose name one knows in connection with his *Dictionary of Practical Medicine,* published in 1882. For many years Copeland's father was tacksman, or tenant-farmer, of Noss's 722 acres.

*   *   *

Lerwick, the county town of Shetland, stands on Lerwick Bay which, together with Bressay Sound, beyond, forms one of the safest and most commodious of Britain's natural harbours. The present town dates only from the early years of the seventeenth century, though much of it would support the assumption of an origin more ancient. By an Act of 1624 'anent the demolishing of houssis of Lerwick,' which the sheriff of Orkney and Shetland ordered, because of the great wickedness of the inhabitants, and of the Dutch seaman resorting there in pursuit of the profitable herring, much of it was laid waste. An interesting echo of the Dutch who frequented these waters at that period is to be heard in the contemporaneous tombstone erected on Bressay in 1636 to the memory of the Dutch commander who, when in pursuit of Portuguese slave-traders, took shelter off that island, and died the following day. Local tradition has it that, winter and summer, a small cavity at the lower end of this recumbent stone always contains at least *some* water.

This tombstone is one of three relics of the kind to be found within the fragmentary ruins of St. Mary's Church, which adjoins the site of an ancient broch at Cullingsburgh, in the north-east of Bressay. It consists of a slab of blue slate 5½ feet long, and a little over 2 feet wide. The following is a translation of its Dutch inscription: *Here lies buried the brave commander, Claes Jansen Bruyn, of Durgendam, who died in the service of the Dutch East India Company, August, 27th, 1635.*

The previous year the Dutchman thus commemorated had been in command of a squadron engaged off the Mozambique coast in chasing a number of Portuguese slave-trading galleons. On February, 9th, 1636, he left Surat for home, aboard his ship, *Amboina.* Contrary winds and other misfortunes, including the loss through disease of twenty-nine of her crew, greviously delayed the homeward voyage. By the time she reached Bressay Sound on August 26th, many of the survivors aboard her were already weakened by illness. The following day her commander died there; and so on Bressay he was buried. The crew's condition necessitated the *Amboina's* lying off Bressay for several weeks. Not until October, 10th, did she reach Texel with her cargo of Persian silk.

Records show that at this time a fleet of 2,000 Dutch fishing-vessels assembled here during the summer fishings. Subsequent warfare with Holland was to bring some of Cromwell's naval vessels into Lerwick waters; and it was at Cromwell's command that, in order to keep the Dutch in their place and to let them see that even the distant Shetlands were part of the Great Britain he would defend against them, a little fortress was built at Lerwick. The premature dismantling of this fortress enabled the Dutch, in 1673, to render it useless, and at the same time to devastate much of the town.

Roughly a century later the present fort was built, and named Fort Charlotte—which reminds me of a story told about the late William Adamson, a Secretary of State for Scotland in Ramsay MacDonald's Labour administration, and an ignorant sort of chap whom I knew. Adamson, a Fifer, was proud of his fluency in Braid Scots. When on an official visit to Lerwick, he was conducted round the town's places of interest.

'Wha biggit this place?' he asked, upon his arrival at the arched gateway to Fort Charlotte.

'Oliver Cromwell,' replied his escort.

'God bless me!' responded the astonished Adamson. 'Did yon heid-strang deevil get this faur?'

Today Lerwick is one of the most important herring ports in the country. Every possible activity connected with the herring industry is pursued there. To the town's considerable export trade in cured herrings should be added that in knitted garments, in cattle, sheep, and Shetland ponies.

The Shetlands, of course, are famed for their ponies—for their shelties. It was on Bressay that the Marquess of Londonderry, many years ago, had a pony farm from which unfortunate consignments of these creatures went regularly to the coal-mines. The mares were kept on Bressay, and the stallions on Noss. These ponies had been bred with a view to maintaining their diminutive proportions. John Tudor, in his monumental work on *The Orkneys and Shetland*, published in 1883, tells us that there used to be held at Seaham, in County Durham, an annual sale of such shelties as were considered too large or too good for pit-work. At one sale of the kind, held in 1878, thirty lots of these 'horse-ponies' realised an average of £25 apiece. Writing of the shelties in his *Tour*, a century earlier, Low relates that they could then be had at prices ranging from twenty to fifty shillings each, and that a few years previously the cheapest cost was but five shillings, and the dearest no more than twenty.

* * *

Throughout breakfast at the Queen's the following morning, as if by the design of somebody concerned that I should appreciate the mental climate of the Isles among which I now found myself so agreeably situated, there was propped up in front of me a fragment of folklore describing Menia's and Fenia's grinding of King Frodi's mill:

> *King Frodi possessed a mill which gave forth anything the grinder wished for, but no one in his own country had strength enough to turn it. Travelling in a strange land, he one day saw two women of powerful stature, ploughing. He asked them if they would attempt to turn his mill, and they, being possessed of supernatural powers which he did not know of, consented. They accompanied him to his country, and were successful in turning the mill, which produced gold and riches of every description. Menia and Fenia at last tired of their task, and asked leave to return to their own country; but King Frodi refused, and would not let them stop grinding, even to take food. The women brought their supernatural powers to their aid, and summoned their countrymen, who destroyed Frodi and all his Kingdom.*

This specimen of Shetland folklore induced in me a receptive state of mind. It sent me forth early upon the flagged thorough-fares and cobbled lanes of Lerwick that morning, intent upon absorbing what local colour I could before embarking for Bressay. The exhibits in the small museum housed in the council chamber at the Town Hall were a sufficient inauguration in this direction.

There I saw my first looder-horn, or fog-horn, such as was sounded throughout the Shetlands in olden days to announce to fisher-folk families the safe return to port of their craft. The looder-horn, a large ox-horn, was fitted with a detachable mouth-piece. Each family is said to have been able to distinguish the sound of its own particular looder, and was thus made aware in the dark hours of the return of its own boat.

No less interesting was the model of a sixern, the open-decked boat the Shetlanders of old rowed and sailed out to sea, so called because it had a crew of six, each of whom had his own kabe, or wooden rowlock. Attached to the gunwale near each kabe hung the loop of rope known as the humbliband. This kept the oar in place at the kabe—a much better arrangement, the Shetlanders held, than pins and ordinary rowlocks, since it ensured the retention of the oar when the rower in heavy seas caught a crab or missed a stroke. The model I mention is complete even to the miniature blaand keg which held the sour milk drink when at sea. Keg and biscuit-chest contained all the drink and food with which the sixern's crew sailed. The chest also held each man's mug, the ship's kettle, and the looder-horn. Elsewhere aboard were stowed such utensils as the skup and the ouskery. The former was used for baling the sixern: with the latter the herring catch was scooped up to be poured into boxes or baskets.

The Shetlanders are, indeed, proud of this model. It was made by John Shewan, an old boat-builder. John was the last of Shet-land's old craftsmen, certainly the last to build at Lerwick a sixern, and in conformity with the traditional dimensions— roughly 72 feet overall, $5\frac{1}{2}$ feet at its deepest, and 20 feet at the beam. It carried a single lugsail. What makes this model of particular interest is that the only other known to have existed in this country was destroyed at Liverpool in a blitz on that city.

The sixern belongs to the days when the seamen of these islands, having no compass, relied on the Seventh Wave to bear them safely homewards. The Seventh Wave, they say in Shetland, always makes for land. The older fishermen (not the younger, who have known only such boats as are fitted with compasses, and are motor- or steam-driven) still speak of the *Midder Di*, the Mother Wave, just as they do of the huggistaff when alluding to the gaff on which big fish are hooked to be dragged aboard.

We mustn't tarry much longer in Lerwick's museum for fear
of missing the ferry to Bressay, and thus arriving too late, even
in midsummer, to see, before sundown, the bird-haunted cliffs of
Noss. But one might just mention its collection of flints and
arrowheads, its ancient utensils in stone and bone, the quern-
stones, and the stone basins in which grain was crushed with a
heavy, wooden mallet, and the cruisies, known to the Shetlanders
as collie-lamps. Then there are some specimens of the tiny
spinning-wheels on which the island womenfolk spin the wool of
their own sheep for their celebrated Shetland shawls. So finely
spun are these, they boast, that they can be pulled through a
wedding-ring.

Archaeology has long been a serious preoccupation in the
Northern Isles; and now that the finds made locally are not
immediately sent, as formerly, to the National Museum of Antiquities at Edinburgh (such as that sculptured antiquity bearing
Ogham inscriptions, and known as the Bressay Stone), the
number of exhibits in Shetland's own museum is steadily
increasing.

*   *   *

Intent upon reaching Noss and the gap in its cliff-edge from
which, as I had been told, the island's gannet colony might be
photographed at eye-level and at close quarters, with the morning
sun illuminating its nesting members, I now hastened through
Lerwick to catch the first ferryboat returning to the intervening
island of Bressay after delivering Bressay's daily contribution to
Lerwick's milk supply. Weather conditions were unpromising;
and it was disheartening to learn from Mr. Mainland, upon my
disembarking there, that a landing on Noss could not be effected
until the wind veered round a bit to the west, and that, even
when it did so, the shepherd, Noss's solitary human inhabitant,
might feel disinclined to venture across Noss Sound, since the
preceding night's gale must have whipped up seas which would
take several hours to abate. 'He wouldn't come out for you on a
day like this,' Mr. Mainland assured me; 'and, even if tomorrow's
a good day, you won't get over if the wind's at all in the south.'

If my trip across the much wider sound to Bressay was un-
eventful, my journey across that island certainly wasn't. It began

propitiously enough by my finding in my path—literally, at my feet—the nests of eider-duck. To the Shetland folk these downy creatures are known as dunters. Their eggs are so often sought by collectors living remote from such nesting-sites that quite a little trade in them is carried on by the older school children on Bressay. The children frequently find rare birds' nests on their island. From these they blow and sell the eggs. They gather them, within reason, at the beginning of the nesting-season, leaving the nests undisturbed thereafter, so that each year the birds may hatch their usual quota.

No one reasonably observant could be abroad for more than a few minutes on the moors of Bressay early in June without locating the nests of eider-duck, each with its clutch of four or five eggs covered over by a dark down so beautiful to the sight, and so soft to the most sensitive touch as to be impalpable.

Not without interest to the ornithologist are those rectangular, drystone enclosures of but a few square yards, scattered everywhere about the hillsides and moorlands of the Shetland Isles, often some distance from human habitations, and by no means scarce on Bressay itself. These resemble the sheep-fanks of the Highlands and Western Islands. They are, however, the Shetlanders' 'plantie-crubs', within which, as a protection from biting wind and nibbling beast, they rear, between June and August, the cabbage seedlings transplanted in March or April to their fields or back-gardens. These cabbage crops are raised primarily for the feeding of livestock, especially for the sheep, during the winter months. The Shetlanders find them more dependable than the neep crops which sometimes fail them.

To the ornithologist pursuing his observations in the tree-less and almost hedge-less Shetlands, the 'plantie-crubs' are of some special significance, since they afford nesting-places for a variety of land-birds. Starlings and blackbirds use them regularly. So also do several small species—wrens, for example. Though the primary purpose of the 'crubs' is the rearing of cabbage seedlings, one finds all manner of interesting growth in them—'and here and there,' as Richard Perry wrote, 'a sudden beauty of garden flowers, with flaming tiger-lilies, grotesque Australian daisies, mimulus from America, and English lupins and michaelmas daisies growing side by side with gigantic, cultivated strains of lady's-mantle and rose-root, whose whorled clumps are such a feature of the Shetland cliffs.'

*   *   *

Having been assured that the summit of the Ward of Bressay, the island's highest point, rising to a height of over 700 feet, provided one of the most extensive panoramas in the Shetland Isles, I felt it incumbent upon me to ascend it on my way across the island toward the Sound of Noss, even though this entailed a detour of three or four miles. For this I was well repaid by the panorama it afforded. To the north lay Yell, Fetlar, and Unst: to the north-east the Out Skerries. To the south-west towered Sumburgh Head, in proximity to which I had landed with high hopes the previous day. Beyond the low-lying territory occupied by the airfield, one glimpsed the Fair Isle. To the west, beyond hilly and peaty Mainland, one discerned the jagged skyline of Foula, where, in the nineteen-thirties, was shot *The Edge of the World*, a film inspired by, and based upon, one of my early books, which also supplied its title.

When eventually I reached the grass-grown track falling steeply through pastures thick with ewes and their newly-born lambs, Noss and the intervening channel, impressive in the sunlight under a fine range of billowy clouds, suddenly came into view; and I could not but recall that awful traffic in pit-ponies as my eye scanned the pastures upon which those hapless victims of man's callousness had once roamed. Standing by the fringe of the sound immediately opposite the shepherd's house, now the only habitable place on Noss, I shouted loudly in the hope of attracting attention. But nobody responded—not then, at any rate. The shepherd, I thought, must be on his rounds, or perhaps on a day's absence from the island. What better could I do in the circumstances than climb Ander Hill, rising not too precipitously behind me? I might at least discover up there a spot from which Noss, in its entirety, could be photographed. With this in prospect I was about to turn my back to the Sound when I fancied that I had detected on Noss the merest evidence of human movement. I paused to watch intently. To be sure, a man was standing motionlessly by the gable of the house. The shepherd himself, without a doubt! Determined that he should see me and meet my immediate requirement, I whistled shrilly, and at the same time vigorously waved aloft the extended legs of my camera tripod. The figure moved hesitatingly, and then descended to the shore, close at hand. It *was* the shepherd. He scanned the narrows rolling obstreperously between us before a southerly wind, the only wind rendering this crossing difficult, if not impracticable. As he examined me at this distance, I perceived that he was expecting somebody else. Launching his rowing-boat, in the matter of a few minutes he came within speaking range.

'Are you wanting across?' he asked, while deftly turning his craft about so that he might back into the sheltered creek at which I stood. This enabled me to step aboard dryshod and with ease.

'Yes, indeed!' I answered.

'When I saw you waving on the hill there, I thought you were Mr. Perry.'

'Mr. Perry!' I repeated. '*Which* Mr. Perry? Not by any chance *Richard* Perry, who writes books about islands and birds?'

'Yes, the very same!' the shepherd responded. 'That's the one all right. An' he's done a book on Noss too. It's no' published yet. He lived here wi' me for months last year, studying the birds and everything about the island. Oh he'll be back here the day yet! That's the way I thought you were Mr. Perry when I saw you on the hill over yonder, waving away. I saw you all right.'

'I hope you're not disappointed at finding I'm somebody else!' I interposed.

'Well, not just exactly,' he replied; 'but I like fine when Mr. Perry comes over. He's a grand sort o' chap. I like fine tae hae a crack wi' him.'

The shepherd scrutinised my person closely ere he allowed his eye to alight on my camera and its appurtenances.

'There' was a gentleman here last year wi' a thing like that,' he said, taking a hand off an oar in order to point to my tripod. 'He wanted to photograph the bonxies. An' will you be a birdie chap yoursel' by any chance?'

In answering him, I now had to confess to the slenderness of my ornithological knowledge, though, as I now told him, I *was* familiar with Mr. Perry's books, and had even quoted in one of my own a telling passage on cruelty to animals from *I went a' Shepherding,* then recently published. This inspired in the shepherd that flicker of interest in me and my doings which, as we stepped ashore on Noss and began to haul the boat beyond the reach of a rising tide, prompted his asking me whether I, too, had come to this isle on account of its bird-life, and whether I appreciated that it was now scheduled as a bird sanctuary. I answered that, although I had never landed on Noss before, I was reasonably conversant with its attractions for anyone interested in field pursuits of a humane nature, and especially in bird photography.

My enthusiasm where such photography was concerned had already been whetted by what I had seen in Lerwick. One would have difficulty in entering a home in this northern metropolis without seeing such magnificent specimens of it as to make one feel the futility of even attempting to emulate them. Lerwick people have made a serious study of the birds of Shetland; and several have spent years in photographing them. Nowhere have I seen finer bird photography than that in which some of them have specialised. Their photographs of the Arctic and the great skua (birds to which we shall refer shortly) are as perfect as anything of their kind in existence. In pursuits such as this, the inhabitants of the Northern Isles differ from those of the Western, where scarcely anyone except lightkeepers isolated for months on remote rock stations display any interest at all.

*　　*　　*

From the shepherd's threshold the Noup of Noss lay at a deceptive distance, albeit to the eye the intervening moorland appeared to rise not too steeply on an inclined plane terminating, as I eventually discovered, in an old dyke fringing the abrupt edge of the cliffs at their maximum altitude. This dyke was built long ago to restrain livestock from venturing too near an edge providing but treacherous foothold. From my experience of such terrain in various parts of the world, I knew that the going would be rough and exacting even though, at this range, the extremely broken nature of its surface, so beset with pitfalls, could hardly be detected with the naked eye. Nobody scanning this moorland for the first time from the shepherd's doorstep could have imagined how difficult to traverse are those forlorn stretches where the rains of centuries have washed away so much of the peat-hags as to have created a scene resembling the no-man's-land we knew in Flanders and on the Somme in our teens.

'O yes! Mr. Perry'll be here the day yet!' the shepherd assured me as I set off for the Noup, anxious to see, ere the sun had travelled much farther, the myriad seabirds nesting on its cliff-face. 'See now that you're back by four, when there'll be plenty of water for me to ferry you back to Bressay!' he admonished as I picked up my camera, ready to depart. 'Mr. Perry's sure to be here by then.'

Fortified by the pleasant prospect of meeting this notable naturalist on my return from the Noup, and also by the shepherd's promise of as much tea as I could drink (tea with milk, I realised, having noticed that, besides the sheep and the lambs and the shepherd's two collies, there were on the island at the time two cows in milk), I went forth to fulfil the purpose which had brought me hither so expeditiously from Barra's Cockle Shore. I saw very little prospect of my being back by four, having landed on Noss some hours later than I had planned when embarking at Lerwick. Yet, I knew from long and intimate acquaintance with regions of this kind that, even although the sound was only 200 yards or so wide at the customary ferrying-place, this comparatively narrow strip of water, in the matter of a few minutes, could become unnavigable for any craft whatsoever. Thus Noss is sometimes completely cut off for a week or more.

Nobody lives and moves and has his being among such exposed islands without an eye on the weather. Vigilance in this regard is essential to personal safety. Nowhere among the 460,000 acres of bog and moorland and rough pasture comprising the surface of the Shetland Isles is one ever more than three miles from the sea. There is scarcely a mound or hillock, let alone a hill, from which the sea is not visible, or its sound inaudible as it sweeps in and out of the innumerable voes forever penetrating farther and farther inland, seeking inexorably to divide and conquer. Who could exist in so precarious an environment without being wind- and tide-conscious?

*   *   *

On my long climb among nesting eider-duck, through the mossy, peaty, and heathery interior of the island, I unexpectedly

entered territory held exclusively by the bonxies, as the Shetlanders call the great skuas—'the handsomest, pluckiest, cheekiest,
and most devil-may-care of all sea-fowl, if not of all birds,' says
Tudor. Here, on every hand, these belligerents were nesting. The
very broken ground around me was infested with them. Of their
audacity, there was certainly no lack of evidence. My traversing
this nesting-ground of theirs so infuriated them that they mobbed
me *en masse*, and then proceeded to dive-bomb me in relays,
without intermission, until they had conducted me well beyond
their boundary. To have a dozen of these mighty birds scream resentment overhead, and then suddenly swoop down to within a few
inches, as if intent on decapitating one with the sweep of a wing
menacingly adjusted, can indeed be terrifying. The draught their
great, powerful white-banded pinions create about one's ears
when zooming within striking distance, thoroughly dishevelling
the hair of one's hatless head, is truly alarming—so much so, in
fact, that I felt I couldn't continue my uphill journey to the cliffs
with any composure until I had armed myself against attack by
extending fully my tripod legs to carry them upright above my
head like a mast. Though this scarcely lessened their attentions,
it did mean that they paid them a few feet higher up, and therefore less dangerously. When a wing-beat, striking the tripod legs
held firmly aloft in the manner indicated, knocked them sideways,
one promptly brought them to the vertical position again, fearing
lest even the most temporary dislocation of this defence should
concede the attacker the opportunity for thrusting a powerful
beak into one's uncovered cranium. Had I known beforehand that
this particular species of skua, having just hatched its young,
sometimes swoops low enough to enable it to deal one a resounding
kick on the head with its large, webbed feet, I would have tres-
passed less incautiously on its territory. But for my fending
tripod, one or two unusually aggressive pairs might well have
kicked me senseless as I squatted to examine at close quarters their
fawny nestlings, almost completely hidden under a smothering of
down. Even in babyhood, as I soon discovered, the great skua
demonstrates his natural ferocity. Long before he can stand on
his pale blue legs and feet, he will strive to rend the most
innocent hand stretched toward him.

This intrusion of mine upon our northern skuas recalled

Herbert Ponting's account of their southern cousins' behaviour when likewise in defence of their territory. There is nothing refined either about the male or the female skua, Ponting wrote. Both are scamps and malefactors. 'They would fly towards us from the rear and, carefully making allowance for speed and distance, discharge a nauseating shower of filth.' More than once, Ponting himself was the recipient of this discourtesy.

If my experience with Shetland's bonxies smacked of danger, it also conveyed something of exhilaration; and so confident did I begin to feel in the protective virtue of my tripod borne aloft that, on returning from the cliffs' edge an hour or two later, I deliberately invaded their domain again in an endeavour to photograph a bird or two, and perhaps a nest, although I realised that, unaccompanied, this could hardly be done satisfactorily. The birds' threatening behaviour made it difficult to concentrate on the niceties of aperture and timing. If I had had with me a companion, things would have been easier. He could have distracted or deflected those waves of bombing birds, leaving me freer to adjust the camera.

The bonxies were determined to press home the attack in order to arrest my progress at their nesting-ground; and I fancy they thought they had succeeded when, for a minute or two, I stood stock-still to observe them the more closely. Many now settled on their own particular mounds. These resembled tiny, green oases in a wilderness of heather and bog-land. Generally circular, and three or four feet in diameter, they had the appearance of having been used as nesting-sites year after year. Having originated as small protuberances, they had been added to during each breeding-season, each successive year's accretions being firmly trodden down by the powerful feet of these heavy birds. Thus, what once had been small, inconspicuous eminences were now the most prominent feature of the landscape. Scattered round them were the feathers and dismembered wings of the birds upon which these ruthless predators live—mostly kittiwakes, which breed on Noss in their thousands. Day in, day out, the bonxies slaughter the nesting kittiwakes, and also herring and greater black-backed gulls, often chasing and harrying the latter until they compel them to disgorge. The bonxies sometimes even succeed in forcing gannets on the wing to spew up fish.

This they sweep down to catch ere it falls into the sea.

What a murderous business all this is! Yet, I find it a little difficult to sympathise with the black-backs, having witnessed the unceasing warfare they wage upon those innocent wonder-birds, the Manx shearwaters. But what chance against the bonxies —against the great skuas—have kittiwake and guillemot fledglings planing for the first time to the sea from the particular ledges of cliffs their forebears have colonised since time primeval? Richard Perry, who has made an intimate study of Noss's bird-life, has estimated that every night during the month of July, when the bonxies would appear to be more savage than at any other season, they take a heavy toll of the young guillemots, and in so doing litter the sea at the base of the cliffs with countless carcases. The only birds on Noss which the bonxies are unable to persecute to the full are the powerful gannets, some thousands of which also nest on the Noup's cliffs.

Tens of thousands of guillemots breed on its cliff-ledges, laying their solitary egg, not in a nest, but on the bare rock. Incubation, which takes roughly a month, is shared by both parent birds; and, since the ledges are seldom level, both have to exercise the greatest care, especially when relieving one another, lest the egg, so precariously poised, rolls off into the sea. The slightest clumsiness on a parent's part may precipitate the egg into the seething ocean below. I have watched nesting guillemots changing over very gingerly on the tilted cliff-ledges among the remoter islands of Scotland, and have observed how acutely conscious they are of their eggs' instability. How ready they are, furthermore, to counteract an egg's movement whenever it threatens to respond to the influence of gravity! And how defenceless against the predatory skuas are these birds, which come to land only to nest, there to contend with a world pitilessly hostile.

In the early days of July such guillemots' eggs as have survived mishap of one kind or another produce chicks which remain at their birthplace for three or four weeks, during which time they are fed with fish caught roughly thirty miles from Noss by the parent birds. By the end of July the young are beginning to show an interest in the world around them, moving a little unsteadily upon their rocky habitat, stretching and flapping their immature wings, and craning their necks to look down some hundreds of feet at the ocean, the home to which, any day now, they will descend.

Among the most wonderful and sobering sights I have ever witnessed has been that of young seabirds hesitating on the brink between earth and sea, and then, in response to some inward and impelling impulse, planing to the ocean for the first time, just as their ancestors have done for aeons. The young guillemots usually make their descent as night falls. Their chances of not being slaughtered on the way down by skuas and greater black-backs depend largely on whether each is accompanied by one or other, or by both, of its parents, who will fend off, as best they can, any devouring pursuer. Here let me quote a pertinent passage from the writings of my friend, John Peterson, who knows Noss and its bird-life intimately, and who has photographed the birds of his native Shetland with enviable distinction:

'Immediately parent and chick reach the water, they make for the open sea, for there lies their greatest safety. At times the water at the foot of the cliff is crowded by adult birds clamouring in a chorus which seems to grow and fade in a wild, urgent cadence, and becomes at times almost an incantation. At first it might appear to be parent birds calling on their offspring to join them on the water: the amazingly shrill piping of some young bird which has fluttered down, unaccompanied, and is without protection, seems to receive a heightening volume of response; but, unless its own parent is there, its overtures to individual birds are met with cold indifference, if not with active repulse. More often than not, it is only a matter of seconds before some hungry bird swoops down and devours it. . . . In a few days it is all over. The last young guillemot of the colony has launched itself from cliff to sea, and the last adult has abandoned the nesting ledges. Their interest in land has grown dim: the many thousands of erect little figures, which jostled one another so incontinently for a foothold on the narrow pavements of their vertical city, are dispersing out into the North Sea and the Atlantic. There winter, with its gales and darkness, awaits them; but their only thought is to escape

the land. Not until the days begin to lengthen will they think
of returning. Then, moved by the urge to maintain their
species, they will venture again to land, a multitude, which,
if it is to be judged by human standards as half comical, half
stupid, must also be accorded its place among the most
intrepid voyagers of the wide ocean spaces.'

*   *   *

But I want to return for a moment to the great skuas—to the
bonxies—since these powerful predators are the most unique of
the larger birds found on Noss. I should mention that they also
breed in considerable numbers at Hermaness, the northern peninsula of Unst, another of the Shetland Isles. Some years ago, the
Nature Conservancy declared Hermaness and Noss national
nature reserves. An extension of the former by some 1,300 acres,
announced in the autumn of 1958, brought its total area up to
roughly 2,400 acres. Hermaness and Noss were declared nature
reserves simultaneously. The former embraces an area of slightly
more than a thousand acres. This includes two islets lying offshore, Muckle Flugga and the Out Stack, both already referred to.

In 1850 there were only two pairs of bonxies on Hermaness.
For several decades their numbers were kept down there by the
profitable activities of collectors. When, roughly a century later,
Hermaness became a nature reserve, its skua population was put at
500 breeding pairs. This striking increase was attributable princi-
pally to the protection vouchsafed the species by the Edmonston
family, its proprietors, and later by the Royal Society for the
Protection of Birds.

Both species of skua, the great and the Arctic, or Richardson's,
are also found on Foula and on the Fair Isle. The Arctic variety,
with its protruding tail-feathers, is a bonnier bird than the
bonxie, and certainly a more graceful flier. The inhabitants of
these northern islands call him the Scootie Allan. Precisely why,
I cannot tell, although I have been informed that the name is in
some way connected with his practice of so scaring gulls on the
wing as to compel them to vomit. What the gulls thus disgorge,
the Arctic skua deftly seizes in mid-air. On Noss this bird nests at a
lower altitude than does the bonxie. It has been known to do so
as low down as the old dykes in the vicinity of the shepherd's

house, and therefore at scarcely any height above the Sound of Noss.

Among the more interesting ornithological records I was privileged to see when visiting the Fair Isle Bird Observatory a few years ago were those relating to bird-ringing recoveries. Shortly before this, the following picturesque letter was received there from Bermeo, in the Basque province of Vizcaya: 'We have the honour to inform you that a month ago a fisherman from this harbour have catched at quay wave-breaker a black marine bird with an aluminium ring. We mourn not knowing the name of this bird.'

Since the ring bore the number, 406/929, there was no difficulty in identifying the bearer of it as a bonxie ringed when a chick on the Ward Hill of the Fair Isle on July, 18th, 1951. An Arctic skua chick ringed at Swey North, on that same island, on July, 10th, 1952, and recorded as having been on the wing on the 31st of that month, was shot at Benguela, in Portuguese West Africa, on October, 25th, 1953. Such records would seem to show that Shetland's skuas travel between 6,000 and 7,000 miles to and from the territories where they winter.

*   *   *

The colony of great skuas on Foula is probably the largest in the northern hemisphere. During the last few years these ruthless killers have been extending their range to the Orkneys, nesting freely upon the higher reaches of some of the less accessible isles. A year or two ago, a correspondent, in testifying to their belligerent and voracious nature, observed a big, brown bird with white wing patches—patently, a bonxie—attack and kill a fully-grown eider in the Bay of Firth, a large sheet of water situated about the centre of the Orkney group of islands, where several pairs of eiders were quietly feeding with their broods in sheltered baylets close inshore. This same intruder then swooped down upon an eider-drake, separating him from the others of his kind. The drake dived to avoid it. Whenever he re-appeared, the bonxie again descended upon him, repeating this tactic several times, and thus denying the victim an opportunity of drawing breath. Soon the exhausted drake lay helpless on the surface of the water, its wings extended, its head down. It was now completely at the mercy of its attacker, who forthwith proceeded to devour it.

The problem confronting ornithologists at the present time, where such northern regions as Noss and Hermaness are concerned, is the manner in which the protection afforded the multiplying bonxies threatens the several species on which they prey so rapaciously.[1]

The other birds breeding in the remoteness and scenic grandeur of Hermaness, elsewhere than on its cliffs and islets, are similar to those found on Noss. They include the almost equally dramatic Arctic skua, the red-throated diver, and the eider-duck. On Hermaness's cliffs and islets one finds thriving colonies of guillemots, razor-bills, kittiwakes, stormy petrels, gannets, and, of course, puffins, red-nebbed like bibulous old gentlemen in yellow boots several sizes too large for them as they trail them astern when on the wing. Gannets began to colonise both Hermaness and Noss during the First World War. They have done so with remarkable rapidity and success. Some 4,000 pairs now nest on the former, and almost as many on the latter. These gannetries are still expanding.

Since the immense bird-haunted cliff-face at Noss Head faces east, most of it lies in shadow for the greater part of the day. My consulting the map had made me aware of this from the outset; but sundry distractions, as you will have seen, delayed my arrival there. For this, in one way, I was now a little sorry, since cliffs in the shade can seldom be photographed satisfactorily. On

[1] No less cruel and destructive of life is the Antarctic skua, described by Robert Ardrey in his overwhelming volume, *The Territorial Imperative* (Collins, 1967), as the world's most disreputable bird. I first learnt of him and his rapacious ways from celebrated friends, now dead, who had accompanied Scott to Antarctica—from Griffith Taylor and Frank Debenham. 'The Antarctic skua,' writes Ardrey, 'like the great-crested grebe, is a bird of old origins, descended probably from the common stock of plovers and gulls and terns and guillemots. He is a giant, with a wingspread of almost five feet. Like the albatross, he takes a long time to grow up—five years—pairs for life, is monogamous, and returns again and again to the same breeding colony where he was hatched. So far as I know, he has but one charming way. When the polar winter closes down and the breeding season ends, the skuas disperse all about the rim of the Antarctic continent to follow the icepack and live off marine life. Pairs break up, and he may feed in McMurdo Sound, she a continent away. But when October comes, and the Antarctic springtime, both return to the area of the colony. If both have survived, then he finds her or she finds him, and domestic life is resumed just where they left it when the great dark fell.'

the other hand, the slowness with which I ultimately achieved my objective had brought me in almost too intimate contact with a bird about which I had heard much, and of whose behaviour during the nesting-season I was now to be in a position to say something from personal observation. Yet, I did regret my not having reached the Noup earlier in the day with a view to obtaining not only some photographs of the great expanse of sitting birds than was now possible, but also with the certainty that one could have photographed with ease its gannet colony. A small promontory at an altitude of about 300 feet, where surface erosion has left in the comparatively soft sandstone cliff-edge an appreciable gap, enables one to approach in relative safety and comfort within fifteen yards or so of these gannets. In fact, a level platform of a few square yards at this point would seem to have been devised by thoughtful Nature with a view to easing the task of the photographer desirous of erecting his tripod just at the spot from which the gannets can be seen to advantage. Few vertical gannetries in the world, I should think, can be so conveniently situated for photographic purposes. It was at this spot that Richard Perry carried out, over a period of several months, his intensive observations of the gannets, watching vigilantly there for hours on end, by day and by night.

The gannet begins nesting on Noss about March. Its solitary egg is hatched any time from May onwards. For ten to thirteen weeks thereafter, the young birds remain at their cliffy abode, facing inwards, as do all birds resting on the ledges and in the crevices of the rocks. A fortnight or so before leaving their nests, they begin to stretch their wings at intervals. It is thought that, during their last week or ten days ashore, the parent birds do not feed them, and that this may have something to do with their deciding to quit the nesting-place. They make a good first landing, often planing a distance of a thousand yards to the sea. For several days thereafter, they remain on the water, too heavy to rise from it. By the time they *can* do so, they may well have drifted considerably south or east of their native isle. Richard Perry is my authority for stating that a young gannet has never been seen afloat on Shetland waters.

The parent birds, seemingly unconcerned about either the whereabouts or the safety of their offspring, remain seated or standing about the same old nests until the first or second week in October, when they again disperse throughout the Atlantic. The earliest of them to return to the cliffs—probably to the same nesting-sites—do so late in January or early in February, spending an hour or two there each day. By March, most of them are beginning to settle down in preparation for nesting again, adding from time to time to the foundations of previous years' nests a grass-blade or two, or some scurvy-grass, or a bit of dried tangle, or perhaps a withered stem of angelica.

*   *   *

In the south-east corner of Noss, at a distance from it of no more than twenty yards, lies the inaccessible Holm of Noss. This consists of an isolated rock with perpendicular sides at least 150 feet high. At a rough guess, I should compute the area of its flat top at half an acre. When I saw it this June day, it was conspicuous a long way off by reason of the whiteness of the sea-campion covering its entire surface. It reminded one of the south-east end of Skokholm, except that in the case of the latter the surface was as pink with sea-thrift as that of the Holm of Noss was white with sea-campion.

On the Holm, as on the sea-thrifty part of Skokholm, gulls (mostly of the greater black-backed variety, which the Shetlanders term swaavies) nest in considerable numbers. Indeed, they were everywhere upon it this June day. By clapping my hands and simultaneously whistling as stridently as I could, I was able to induce multitudes of them to raise their heads above the campion. They exhibited little alarm, however, confident in their knowledge of the Holm's inaccessibility to all but the wing-borne.

Heaven knows when the last human being set foot on the Holm of Noss. Tradition has it that a seventeenth-century rocker scaled its precipitous cliffs for the wager of a cow, dragging behind him, across the intervening chasm, a fair length of rope and a couple of wooden posts. The latter he drove into the ground near the cliff-edge at the top of the Holm. By means of guy-ropes he was able to rig up a ready mode of transport across the chasm in the form of an oblong box. From a point at the edge of the cliff on the Noss side, at an altitude some feet higher than that of the campioned plateau crowning the Holm, the more intrepid thereafter could swing themselves over the chasm, one at a time; but, owing to the difference of levels, the box in which they thus transported themselves always had to be hauled back to the Noss side, whereas it ran freely of its own accord in the opposite direction. They tell one in Shetland that the man whose initial daring made possible this means of reaching the Holm declined, in a spirit of bravado, to return from the Holm by the box, and insisted on doing so the way he had gone. In so attempting, he fell from the cliffs and was killed. Subsequently, the original box was replaced by one large enough to accommodate one man and one sheep. By this means, twelve sheep were placed on the Holm every year, and twelve taken off.

Many years ago, the entire apparatus was dismantled for fear of accidents. Since then the gulls have had the Holm to themselves, occupying it to the full during their nesting season. 'When the cradle at Noss is about to be slung,' wrote Dr. Edmonston in 1809 in his *View of the Zetland Islands,* 'the gulls, aware of the approaching capture of their young, are unremitting in their efforts to carry them off. From the first moment that they observe preparations making to enter the holm, they become noisy and restless, "and chide, exhort, command, or push them off," so that if bad weather delay the arranging of the cradle, but for a few days, scarcely any are left to be taken away.'

This statement, says Tudor, would have seemed incredible, had Edmonston not been regarded as a keen observer and an accurate writer.

*   *   *

Long before I returned to the shepherd's house, the sandwiches with which I had been provided at the Queen's that morning had been consumed, so that I was hungry as well as tired by the time I reached this threshold at least four hours later than that at which the shepherd expected me to turn up. When still some little way off, I spied a fellow smoking a pipe by the gable of the house, and contemplating the sound. I knew he wasn't the shepherd, who would willingly have ferried me back to Bressay at the hour he stated. He was somebody who had set foot on Noss during my absence at the Noup, from which I ought to have returned so much earlier. But how could I have withdrawn from so enchanting a scene merely to suit the convenience of the island's shepherd? I might never manage to land on Noss again. Indeed, I might well tarry in the Shetlands for months ere I was favoured with such weather for my purposes. If my dilatoriness had put the shepherd to any great inconvenience, I must compensate him.

'Richard Perry, I presume!' were the words with which I now approached the pipe-smoker leaning against the gable.

'Richard Perry it is!' he replied. 'But you have an advantage over me. When I arrived, the shepherd told me someone had landed several hours earlier—another author-looking chap who had set off almost immediately for the Noup.'

My omitting to tell the shepherd my name was intentional. I knew that my not doing so would engender speculation. Visitors to Noss were not uncommon during the summer months; but a visitor carrying a camera and a tripod *was*. Would he be photographing the bonxies, or maybe the birds nesting on the cliffs at the Noup? Or was he doing a bit of surveying? The shepherd couldn't quite tell *what* he might be up to!

With a show of profound thankfulness, I now accepted the joint invitation of shepherd and naturalist to enter the house and be seated. I wouldn't like to tell you how many cups of tea they now plied me with, nor the proportion of Richard Perry's sandwiches I voraciously consumed.

After an hour's quiet converse together, Perry and I made a move to quit Noss. Perry was returning to his apartments at Hoversta, on Bressay; and I had to complete thereafter the extra journey to Lerwick. The last ferry from Bressay had crossed some hours earlier; but Mrs. Mainland, Perry's landlady, assured me that her husband would put me over when, about midnight, he returned home from visiting a neighbour. And, sure enough, he did. We raised anchor to the noisy clamour of Arctic terns nesting on a grassy islet offshore. In the nightless midsummer of the Northern Isles—in the midnight glow of Shetland's 'simmer dim'—one could see many of these birds winging hither and thither, disturbed by the chugging of Mr. Mainland's motor-boat as it passed by. In less than twenty minutes, I was back at the Queen's. Resignedly overwhelmed by the observations and adventures of such a day, I had to concede inwardly that there are times when one must accept surrender in face of the unfathomable, the incomprehensible. 'It is fruitless to explain everything in the natural world in terms of selective value and survival necessity,' writes Robert Ardrey in the volume already cited. 'There are times when one can only record what is true, and dissolve in wonder.'

That day's adventures I recalled vividly while reading in the Spring, 1968, issue of *The Countryman* Jo Moran's beautifully written account of his courageous and quite astonishing perform-ance on the Great Wall of Noss the previous year. Inspired by the achievements of native cragsmen while collecting for human consumption elsewhere the eggs and young of seabirds, Jo had long been fascinated by the mighty rampart reaching at 600 feet its highest point in the dizzy precipice of Noss Head. A noted ornithologist's description of the gannet colony at this nature reserve intensified his interest in it. Especially so did his remark that, of course, its nursery terraces were totally inaccessible. 'At the foot of the cliffs,' writes Jo, 'the sea seemed to be perfectly calm, until a few minutes' scrutiny through binoculars revealed a swell rising a clear six feet against tilted slabs, black with algae. There was only one course to pursue: to rope down from the cliff-top to a pre-selected stance just above the wave-washed belt of algae and begin to climb from there.'

This Jo Moran did, photographing on his way up its nesting gannets, kittiwakes, puffins, guillemots, and fulmars. Hitherto, all photographs of the birds nesting on this stupendous rock-face, including my own, had been taken from the dangerous cliff-edge directly above.

No cragsman's nor ornithologist's library is complete that does not include the issue of *The Countryman* containing Jo's own account of this magnificent feat of his.

CHAPTER SIX

Handa and its Birds

THREE MILES of crow-flight in a north-westerly direction from Scourie, a crofting locality in the Eddrachillish parish of Western Sutherland much frequented by anglers, lies the small, proximately circular, and now uninhabited island of Handa, long known to ornithologists on account of the comparative accessibility of the colonies of seabirds nesting multitudinously on its cliffs of Torridonian Sandstone stratified in layers virtually horizontal. It has an area of 766 acres.

It is Handa's accessibility which has afforded ornithologists the opportunity of observing and photographing the feathered inhabitants of these cliffs at quarters closer and less perilous than can, perhaps, be found anywhere else among the innumerable isles and islets of northern and north-western Britain so densely colonised by seabirds. If Handa cannot claim distinction as the breeding-place of any rare species of bird, it can certainly boast a cliff formation providing one with facilities for prolonged and intensive observation, under the most favourable conditions, of all the commoner species of seabirds at their respective breeding stations. Nowhere, for instance, can one study with less inconvenience the nesting habits of razor-bill and guillemot, the two species of auk found so abundantly on Handa.

Touching upon the rarer species of birds, one might mention that the white-tailed sea-eagle nested on Handa a century ago. Today, the island's chief predators are the more ruthless families of gull, which slaughter so mercilessly and indiscriminately its guillemots and kittiwakes.

One might also add that in recent years the peregrine took up residence on Handa.

*　　*　　*

From the mainland of Sutherland, Handa, etymologically the Sandy Isle, is separated by the Sound of Handa, a channel varying in width from a quarter to half a mile. The island is most readily accessible from the tiny township of three or four crofts shown on the Ordnance Survey map as Tarbet. The crossing from Port of Tarbet to its customary landing-place, in the sheltered south-eastern arc of its periphery, is a mile and a bittock.

The sandy bays at Handa's southern end are a delight. The cliffs intervening there are of negligible height, since the island, conforming to the tilt of its Torridonian component, shelves down in this direction almost to sea-level. As one proceeds northward and north-westward, the surface rises to attain its maximum altitude of 406 feet in the Sithean Mor, a modest eminence which, as its Gaelic name signifies, was regarded by the island's inhabitants in olden times as the abode of the Faery Folk.

Immediately beyond this, and at an altitude but slightly lower, the island's awe-inspiring cliffs, comprising an uninterrupted third of its circumference, drop precipitously to the ocean. At their north-west extremity, and detached at a distance of no more than a few yards, stands the wellnigh inaccessible Stack of Handa, a dramatic formation that would assuredly qualify as the *pièce de résistance* of any small island. In May and June this Stack, thronged with nesting seafowl, is as remarkable a spectacle as any along the coasts of Britain, whether it be seen from the lofty and precipitous edge of the parent island itself, or from a small boat brought cautiously into its mystic shadow, some three hundred feet below.

The Stack's seaward face, though diminutive when compared with similar formations I have viewed from sea-level elsewhere— at St. Kilda, at Mingulay, at Berneray, at the Shiants, and, indeed, at the Noup of Noss, in distant Shetland—is impressive even when its prodigious concentrations of various seafowl and their young have left it completely deserted at the close of the breeding season, taking with them that loud and disturbing cacophony which never fails to supplement grandeur with weird-ness, with something truly eldritch.

*   *   *

To reach the Stack of Handa, before the introduction of modern

climbing methods, one would have required either the mighty
stride of the legendary giant, or a degree of rocking skill and
fortitude less common now than when the St. Kildans scaled
infinitely more formidable precipices to collect seabirds and their
eggs, either for immediate consumption, or for salting down for
use during the winter months.

Though one never heard of a native of Handa having set foot on
the Stack, dramatic accounts of such an achievement are extant.
One recalls that supplied by the naturalists, J. A. Harvie-Brown
and T. E. Buckley, in their splendid volume, *A Vertebrate Fauna
of the Outer Hebrides*. Until 1877, the Stack had remained in-
violate. That year, two men and a boy from Uist managed to
reach it at the behest of a proprietor anxious to deal with the
murderous great black-backed gulls' nests occupying entirely its
flat, lofty summit. This was achieved by means of a rope flung
from the nearest cliff-top on Handa itself. The rope became
sufficiently engaged on the Stack to allow of the boy's courageous
crossing, in order to make it secure enough for the men to follow
him. Having fixed firmly the landward end of the rope before
venturing forth on this dizzy escapade, all three were in a position
to return with a degree of confidence, as soon as they had
dispatched what black-backs they could, and destroyed every
egg of the species they were able to find. The black-backs already
had become too numerous and troublesome in these parts.

Recalling this, I could not but wonder whether this onslaught
of so long ago explained how, as far as I was able to see with the
aid of field-glasses, the top of the Stack appeared virtually devoid
of birds. Indeed, it was the only part of it not colonised by
thronging seabirds of one kind or another. Yet, roughly eighty
years previously, the top had been so infested with black-backs
as to have necessitated the hazard entailed in an attempt to
extirpate them from this very spot, and in the manner just
described.

My glasses enabled me to pick up, on the top of the Stack, no
more than four of the species; but, whether they were actually
nesting there, I could not say—I could not be certain. It seemed
unlikely, since these birds are prone to nesting in anything but
isolation from one another. It also surprised me to find that the
black-backs had not again colonised it, and numerously, during

the intervening decades, as they had done at the Holm of Noss, a
very similar structure very similarly situated.

The marauding black-backs on the flat-topped and now inaccessible Holm had been dealt with about the same time, and in
a similar manner. When the rope and cradle equipment enabling
the more daring Shetlanders to reach the Holm decayed, the
knowing black-backs took due cognisance. Discovering that their
ancient fortress had reverted to its pristine inaccessibility, they
proceeded to re-establish themselves there. Today one finds on the
flat top of the Holm of Noss what is probably the greatest
concentration of greater black-backed gulls in all the Shetland
Isles.

Why has this species not returned to Handa to re-colonise the
equally inaccessible summit of its Stack? One day, perhaps, it
will, although it seems reluctant to do so at the present
time.

That at least one human being must have set foot on the Stack's
summit since that 19th-century adventure is shown by the presence
upon it of a slanting stick. The end of this stick was thrust into it
by *somebody*! But by whom? And when? It is very improbable
that it became implanted there by some sort of spear-throwing
exercise, as it were.

In contemplating this stick from the adjoining cliffs, while
myriads of birds are squawking, screaming, screeching, and
yodelling overhead, at one's very feet, and to a depth of three
hundred feet below, one is instantly transported to the realm of
drama. Here one witnesses the truly fantastic. Here one sees the
vision of a boy clinging desperately to a rope precariously
fastened, of a boy ultimately attaining a limited area of *terra firma*
from which he might never have been able to extricate himself,
the devouring gulls waiting to pick his bones. One visualises the
cloud of black-backs already screaming threateningly about his
ears, maddened by an intrusion they never could have imagined
possible.

Tradition in Sutherland asserts that the last raid upon the
Stack's nesting black-backs was made many a year ago by the
crew of a small, fishing vessel from Ness, in northern Lewis. If
this were so, we might well assume that its members were
inured to the peril involved through their fowling expeditions

SHETLAND. A Bressay shepherd in his native setting, with the township of Brough in the background and Ander Hill on right

SHETLAND. The Great Skua, or Bonxie, nesting on Noss

SHETLAND. The Arctic variety, with its protruding tail-feathers, is a much bonnier bird than the Bonxie

The bird-haunted cliffs at the Noup of Noss

SHETLAND. The shepherd's cottage on Noss, now that island's only habitable building

to the gannetry of Sùla Sgeir. From time immemorial such an excursion was made annually from Port of Ness.

* * *

This brings to mind a historic association Handa has with Lewis. Handa is believed to be the isle of Eddrachillish referred to as the Brieve's, or Judge's, Isle, on account of the following circumstance.

Toward the close of the 16th century, the treacherous John Morison, one of the hereditary Brieves of Lewis (known as Brehons in Ireland) had been obliged to flee that island in face of the wrath of his enemies. Sailing from Ness, the seat of his jurisdiction, he took refuge in the Assynt district of Sutherland. There, along with a number of his partisans, he was slain in a bloody encounter with the MacLeods, whose kinsfolk he had wronged most grievously.

When the Brieve's relatives in Lewis learnt of his fate, they set out for Assynt in their galley, intent upon bringing home his corpse for honourable burial. However, contrary winds delayed their return to Lewis. Indeed, they drove the galley to Handa. There it was decided to disembowel the Brieve's body, and to bury his intestines. With the coming of a favourable wind, the galley proceeded safely to Ness.

* * *

Although today Handa is uninhabited, this was not always the case. Over a century ago, seven or eight families lived there, cultivating its none too prosperous crofts. Sharing with the St. Kildans the distinction of having had a Queen and a Parliament, they subsisted mainly on fish, oatmeal, and potatoes. The disastrous Potato Famine of 1845 spelt Handa's doom.

The island's last inhabitants had sailed for America but a few weeks when Charles St. John (the blighter accredited, on the basis of his own writings, with the dubious distinction of having shot, on Loch Assynt, the last ospreys in Sutherland!) arrived on their deserted thresholds in 1848. In his *A Tour in Sutherlandshire* he describes the pathos of the scene. As his boat neared the island, he observed a large, white cat, seated at the extremity of the small promontory nearest the mainland coast, woeful and

D

forlorn. 'I could not help being struck with the attitude of the poor creature,' he wrote, 'as she sat there looking at the sea, and having as disconsolate an air as any deserted damsel. "She is wanting the ferry," was the quaint and not incorrect suggestion of one of our boatmen.'

St. John mentions, furthermore, his having passed several huts on the island, the turf roofs of which, notwithstanding the shortness of time that had elapsed since the inhabitants had quitted them, were already tenanted by innumerable starlings.

The evacuation of Handa was but part of the largely misrepresented epoch known as the Highland Clearances, aggravated at this particular time by the grievous failure of the potato crop. It is true, on the other hand, that landlords now found it highly profitable to convert their properties into sheep-runs; and, so far as Handa is concerned, the Duke of Sutherland is still regarded as the villain in all that occurred there. He is said to have given the islanders a month's notice to quit, having arranged for their transference to Canada. Charles St. John, however, shows him in the rôle of a kind and thoughtful benefactor, and points out his economic problem in relation to such tenants, stating that 'the lower class of inhabitants take but little trouble towards earning their own livelihood. At whatever hour of the day you go into a cottage, you find the whole family idling at home over the peat-fire. The husband appears never to employ himself in any way beyond smoking, taking snuff, or chewing tobacco; the women doing the same, or at the utmost watching the boiling of a pot of potatoes; while the children are nine times out of ten crawling listlessly about or playing with the ashes of the fire.'

The Duke, according to St. John, having failed in every endeavour which philanthropy and reason could have promoted, at last succeeded in convincing Handa's inhabitants of the advantages of emigration, 'and at a great expense sends numbers yearly to Canada'.

*   *   *

Today, the foundations of the old cottages on Handa are none too easily located: so sunken and overgrown have they become. Even the island's graveyard might be missed by anybody landing there without precise knowledge of its whereabouts.

The island's only habitable building—and I use the adjective in a somewhat primitive sense—was the shepherd's dwelling, an erection of more recent date than the buried homesteads occupying a piece of sloping moorland facing Scourie Bay, situated inland approximately a quarter of a mile from the south-east landing-place. This dwelling, this hut, lying in a hollow comparatively sheltered, had a walled sheep-pen adjoining it. Hut and pen were temporarily in use when a shepherd landed on Handa at lambing-time, and again when it was necessary to clip and dip its sheep stock.

Occasionally, naturalists accommodated themselves as best they could in this hut. A vivid description of its interior when the ornithologists, John Ainslie and Robert Atkinson, moved in for eight or ten days in the summer of 1935, may be read in the latter's beautifully written volume, *Island Going*, published by Collins in 1949.

'It was dim inside after the outdoor seaside glare, though the hut was not at all the rude bothy we had expected. It was divided inside by a string hung with the stock of blackened bedding things: the near end by the door stacked to the roof with wooden sheep-troughs, peat, peat digging tools, rabbit wires and indigenous junk—oddments found and kept in case they should come in useful, such as an old electric light bulb; the far end beyond the blankets was furnished and had a fireplace in the end wall. We had intended to do without fire or cooking. The heads of the two beds were ranged on either side of the fireplace, one bedstead of old iron with a mesh mattress, the other of deal, without. This second bed (mine) was peculiar in having only three narrow boards to bridge the space it enclosed, one each for head, hips and feet. Above the beds were cuttings from *Punch* in driftwood frames. *Punch?*—of course, old copies from the Scourie hotel. . . . The rest of the furniture was a wooden table and an iron arm-chair with only the iron parts left. Mainland rubbish, rather than be thrown away, could usefully go to Handa.

'Such light as there was indoors got in through two tiny windows and through a jagged hole in the roof made, so the fishermen had said, by the unexpected going off of a shepherd's muzzle loader. The chimney was topped with a bottomless bucket, wired on, upside down. The flue was simply a hole left in the middle of the wall; later on the view down it to the glowing peat at the bottom was cosy enough.'

*   *   *

Years ago, when my interest in islands was still quite superficial, and I had no experience whatsoever of photography, I landed on Handa. There I spent a long day, gazing in wonderment at its bird-haunted cliffs. Better informed and more purposefully, I returned to these cliffs in June, 1959, when the guest of the late D. M. Reid and his wife, at Stoer House. I had known Douglas and Margaret for a great number of years; and the prospect of my exploring Assynt, with their house as a base, had been held out to me ever since we first became acquainted with one another. Douglas Reid was, of course, Douggie to those with whom he was associated during his years as science master at Harrow.

A reconnaissance dash northward from Stoer in the direction of Scourie one afternoon, with a view to arranging transport to Handa, brought me to Port of Tarbet by a road not so inferior as my map had led me to expect. My car swept along it very comfortably. A number of small boats moored or beached at the Port itself assured me that I had arrived at the place from which Handa was easily accessible, given reasonably fair weather. The entire island of my quest had come into view as I descended to the crofting township of Tarbet, situated within shouting distance of the Port. Here, at Tarbet then, Donald Angus MacLeod and I arranged that, weather permitting, and on payment of a pound, he would land me on Handa from his motor-boat the following day, would leave me there as long as I liked, and would pick me up at an hour to be decided upon later.

Having already agreed the time at which I should embark, I returned from Stoer next morning, and so precisely at the hour we had agreed upon that Donald Angus, later in the day, was to make my punctuality a pretext for anxiety.

In no time I was afoot on Handa, with a whole midsummer's day ahead of me. Having calculated how long I thought it would take me to explore the island, to photograph it, to reach the extremity where the bird-haunted cliffs lay, to observe the various bird colonies, to get the sun just right for photographing the Stack, I suggested the hour at which Donald Angus might return for me. Off he went; and off I went too. But there was so much to see and to do and to photograph. Furthermore, I found the sloping moorland so much more broken than I had expected. It took twice as long to traverse it than I had allowed for. As I soon realised, this meant that Donald Angus would be calling for me three or four hours earlier than I now wanted.

At the island's highest point, where the Ordnance Survey has erected a triangulation station, I paused a moment to resolve my dilemma. Should I make a desperate effort to keep my appointment with the boatman, or should I improve this truly shining hour, and bestow upon him, at the close of day, all manner of excuses for my having failed to do so? Leaning against the Survey's concrete, I allowed conscience to be suitably dulled by academic speculation. 'What will archaeologists say about this erection if they should discover it, say, a thousand years hence?' I asked myself. 'Will it be referred to as conclusive evidence of the primitive surveying methods employed during the 20th century A.D.?'

*   *   *

The hour when I ought to have been at the landing-place found me perched precipitously on the cliff edge facing the Stack, putting every bit of concentration I could muster into my photography, and at the same time disciplining myself in the matter of exercising every possible precaution against accident. With the legs of my tripod splayed none too firmly at the edge of the cliffs, and very much in the way of my feet at a spot so circumscribed as to render perilous even a few inches' movement, I was hardly in a position to think of dashing back to the landing-place. When thus involved, I heard and then noticed, on the calm sea below me, a chugging motor-boat. Suddenly the engine was switched off to allow of its gliding gently from windless sunlight into the windless shadow of the lofty precipices. The entire absence of wind had permitted one to be a little more daring on these precipices than would otherwise have been justifiable. On the other hand, I knew that an unexpected gust from any direction—perhaps, a sudden change in the pressure of air in the chasm yawning at my feet—would have proved fatal.

For a moment I thought this boat had aboard her a number of sightseeing tourists from Scourie, who were being shown the nesting seabirds from the sea, a sight spectacular enough, as I can assure you. Then it occurred to me that, maybe, it was Donald Angus MacLeod's boat. My field-glasses would have resolved any doubt as to the boat's identity, had I not been poised in such a position as to have made access to them positively perilous. I could not have reached the pocket containing them without precipitating myself or my camera, or both, into the ocean, some hundreds of feet below.

I did notice, however, that this boat had aboard her at least three persons. Having released my shutter and folded the tripod, I crept back to safety. Realising that, for the moment at any rate, I was on the skyline as seen from this boat, I thought I would just deliver myself of an informal wave of the hand. Immediately I did so, the engine was re-started, and the boat turned round and sailed away for Port of Tarbet. If its occupants had been wondering whether I, known to have been the only human being abroad on Handa at the time, was still alive and safe, this mild gesture of mine, I thought to myself, would assure them.

*   *   *

Tired but contented, and reasonably confident that in the dark interior of my camera I had bagged something gratifying, I eventually made my way down to the landing-place, where I should have turned up so many hours earlier. There Donald Angus awaited me, now in company with two of his cousins, both of whom bear the name of Donald John MacLeod. Donald Angus explained that, having failed to find me there at the hour agreed upon, he feared that I had met with an accident. He and his cousins had therefore searched the island for me, but in vain. They had shouted their way across it without receiving any response, and had then taken to their boat in order to search the sea under the cliffs from which they believed I must have fallen. This accounted for the position of the boat I had observed while standing so precariously aloft, my feet none too securely placed, and indeed encumbered by the extended legs of a tripod no more firmly sited. My failure to hear the searchers' shoutings, as I now explained, must have been due to the tumultuous noise of the birds with which I was so fully engaged.

When I got back to Port of Tarbet, I handed Donald Angus a cheque for twice the sum we had agreed upon—adequate reward, I had hoped, for the extra trouble I had occasioned him. He was pleased with this; but I soon jaloused that something more was expected of me. The two Donald John MacLeods, he now confided, were expecting some financial recognition as well.

'Say that you had mustered twelve or, perhaps, twenty of your relatives to scour Handa for me, and had come upon me either dead, or so seriously injured as to have necessitated your carrying me to the boat over that very broken ground, would you have expected me to fork out something for each?'

'Well, then,' he answered, 'that would be different.'

'Quite so!' I replied. 'Let us be a little fair if we can't be very logical. Why not divide with your cousins the extra bittie I added to my cheque?'

*   *   *

I need not recapitulate here all one finds on Handa. Subtler pens—pens in the hands of more competent ornithologists—have done so more acceptably than I could. But I might just say that I did spot a pair of red-throated divers, cruising on one of the island's several small lochs. These lochs, for the most part, are attractive enough if one can thole the stench resulting from the congregations of gulls continually attending to their personal ablutions near their edges, and preening themselves thereafter on their noisome banks.

Nowhere can the guillemot be studied to better advantage than from a position overhanging the rock-faces of Handa. At breeding-time this highly excitable bird may be seen in great numbers on the narrow ledges where that island's sandstone cliffs sheer to the sea. Building no nest, it lays on the bare ledge a single egg, large, pear-shaped, and streaked, and stained with reddish-brown blotches. This egg the mother-bird holds on her feet when she is brooding. The chick is fed on fry and sand-eels. I cite Seton Gordon as my authority for adding that, although a parent bird may have flown for food to some fishing-ground a distance of fifty miles or more, it returns to its chick with but one small fish at a time. If that small fish be dropped accidentally even on a ledge and in close proximity to the chick, the parent makes no endeavour to retrieve it, but flies off again to the fishing-ground whence so recently it arrived. To some extent the chick is already fledged when, in response to the encouraging calls of its parents, meanwhile swimming at the base of the cliff immediately below, it eventually throws itself off the ledge to join them, often watched while so doing by murderous black-backed gulls who now swoop upon it as it strives to follow its parents as far seaward as it can swim. Expert at diving though the baby guillemot is from the moment it lands on the sea, it cannot, of course, remain underwater as long as can adult members of the species. As it rises to the surface for breath before again submerging hurriedly, the black-backs may continue their pursuit until, exhausted, it falls a prey to them, having known but a few minutes of life. I do not accept any longer the comforting view that this slaughterous state of affairs, so universal throughout physical life, is part of what the religious plausibly declare to be a *divine* purpose.

Handa's guillemot colony is one of the largest and most compact I have ever seen. The day I landed there, it certainly comprised several thousand breeding pairs, some of them bridled. They nested so densely on the sandstone ledges that, if even one of them moved, a necessity to do likewise seemed to be communicated instantly to hundreds of guillemots in its immediate proximity. It was as though an urgent wave of motion had been started, and must needs make itself felt—work itself out, as it were—as far away from its source as its successive and diminishing undulations could reach.

In the manner in which so casual and innocent a movement initiated so much disturbance among so many birds, thus also did a single guillemot's bowing. Immediately one of the kind began to bob and bow, another did so, and another, and yet another, until hundreds were bobbing and bowing simultaneously, as if actuated by some ingrained tribal ritual peculiar to their own species.

If their antics provided diversion for the eye, their noise made more than ample din, even for the dullest ear. When no longer content to stand—or sit—and stare, they took to the wing in a

The Stack of Handa

HANDA seen across the Sound of
Handa from Port Tarbet

HANDA. The ruins of the old bothy
a grant for the reconditioning
of which was received from the
Howden Trust

The Helena Howden Bothy as
it is after restoration in 1962

vast cloud and with a mighty commotion, screaming and screeching in a manner truly deafening, if not also bewildering. Their
clamour would certainly have terrified—nay, petrified—anybody
who never before had heard such chaos in a setting so isolated,
so impressive, so dramatic.

Any little distraction caused by the intrusion of the most
considerate human observer, even at some distance from these
populous cliffs, or merely the slipping of an egg from ledge or
cranny to break among the serried ranks of vigilant sitters somewhere below, was sufficient to intensify the hubbub almost
immeasurably. You never heard such an uproar as Handa's
guillemots are capable of creating, and on the flimsiest of
pretexts! One retires from it with a sense of relief akin to that
experienced on reaching a sheltered haven after a thoroughly
sound buffeting at sea.

It would be interesting to know how apparatus designed to
record the volume of sound would respond in the midst of so
noisy and raucous a bedlam.

*   *   *

Thanks to the efforts of that well-known ornithologist, George
Waterston, in 1962 an arrangement was made with the new
proprietors of the Scourie Estate whereby Handa became a bird
reserve—the seventh in Scotland to be managed by the Royal
Society for the Protection of Birds.

With the aid of a grant from the Helena Howden Trust, the
old, derelict bothy has been restored. Re-roofed, and with running
water piped from the loch near by, and now consisting of a living
room and bedroom, it is equipped with four sets of two-tiered
bunks, tables and chairs, a calor gas cooker, kitchen sink, crockery,
cutlery, and cooking utensils. A small annexe contains a wash-
basin, a W.C., and a store. Accommodation can only be booked
by members of the Society, and a small daily charge is made to
cover maintenance expenses. Blankets and pillows are supplied;
but intending residents must bring their own food, sleeping-bags,
pillow-slips, and the like. The living-room contains a small library
of reference books, and is kept snug with peat and driftwood,
which visitors are expected to help in collecting, burning in its
big, open hearth. Over the fire-place hangs a framed enlargement

of my photograph of the old, derelict bothy, taken during that memorable visit to the island in 1959.

The Society has appointed as warden of the island Mr. Alasdair Munro, a lobster-fisherman from the crofting township of Tarbet. It was from Alasdair Munro's boat, and with his assistance, that in August, 1969, Hamish MacInnes, with two companions, Douglas Lang and Graeme Hunter, succeeded in scaling the north-east face of Handa's Stack. They were armed with, in MacInnes's own phrase, 'the ironmongery that modern rock climbing demands'; having achieved their object, after a perilous three-hour climb, they found the top covered with flat, spongy grass, but with no sign of any black-backs' nests.

Two years before, Dr. Tom Patey had achieved the same object by a different feat. A 600-ft. length of rope was carried round the cliff top of the bay that half encircles the Stack and then drawn taut over the top of it. By means of this rope Patey then crossed the shortest intervening gap, thus emulating the exploits of the two men and a boy from Uist some ninety years ago.

CHAPTER SEVEN

The Treshnish Isles

IN A north-westerly direction from Staffa's famed columnar cliffs and caves, and at a distance ranging from 4½ to 6½ miles from that isle, lies the remote and wholly uninhabited group of islands, islets, and skerries known as the Treshnish. The merest relics above water of the vast, volcanic sheet that covered so much of what today, among the Inner Hebrides, is occupied by the sea, they form, as does Staffa, part of the Argyllshire parish of Kilninian and Kilmore, situated, as they are, off the west coast of the volcanic island of Mull, from the virtually shoreless cliffs of which parish the islet, Carn Burg Beag, most northerly of the Treshnish, lies a couple of miles. In terms of crow-flight the Bac Beag, most southerly of the group, is about ten miles from the Iona ferry. Running in a south-westerly and north-easterly direction, these isles span, almost in a straight line, a distance of just over six miles.

Except for the occasional landings made by lobster-fishermen operating in the surrounding waters, by those who winter cattle on Lunga, much the largest of the Treshnish, and by those who, during the more clement months of the year, enjoy a day's field excursion to them, they seldom take the weight of human foot.

The six principal islands of the group, enumerated from south to north, are Bac Beag, Bac Mor, Lunga, Fladda, Carn Burg Mor, and Carn Burg Beag. The mile-wide channel separating Lunga and Fladda is thickly strown with islets and skerries, each with its own name commendably recorded on the one-inch Ordnance Survey map.

Some conception of the immensity of the volcanic field of which the Treshnish Isles survive as a visible and virtually insignificant remnant may be obtained on a clear day by the panorama to be had at the highest point of Lunga. Especially is this so when, facing east, one scans Mull's mountainous mass of Tertiary basalts. In that island's splendid Ben More, these basalts, though so denuded through every agency of epigene weathering, operating unceasingly throughout a diuturnity rendering meaningless any human conception of Time, still maintain, at 3,169 feet, their loftiest reach in Scotland. Incidentally, Tertiary territory, terraced as in the case of Lunga and of identical formations on the west of Skye and elsewhere, is good cattle country, although I feel it almost inexcusable that I, forty and more years a vegetarian, should be mentioning this.

In the Treshnish Isles, as elsewhere in this region of the Inner Hebrides, and quite apart from such phenomena as the hexagonal columns seen to such advantage at Staffa and Ulva, the denudation of aeons has carved out of their basalts a rock architecture of a castellated nature, replete with appropriate towers and turrets, bulwarks and bastions. Among the more remarkable examples is Dun Cruit, a rock so called because it resembles in shape a huge Celtic harp, *cruit* being the Gaelic for a harp. Dun Cruit, situated on the west side of Lunga, and separated from it by a narrow channel, is precipitous on every side, challenging even the most intrepid rock-climber, a quite minor crack on its north side keeping alive the hope that one day it may furnish that minimum of foothold essential to a successful ascent.

In common with all such inaccessible stacks throughout our northern and western seas, Dun Cruit is the nesting-place of seafowl. Kittiwakes, fulmars and auks nest numerously upon it. Noisily assembled there at the breeding season, they provide, day and night, and almost without intermission, a music harsher than the gentle chords of the harp, as Seton Gordon puts it. Barnacle geese, frequenting the machar-lands of the Outer Hebrides in their thousands, also visit the Treshnish Isles. Likewise do petrels. The richness of Lunga's grass ensures not only ample winter feeding for cattle or sheep pastured there, but also wintering for considerable numbers of blackbirds and starlings. In the winter-time the green grass covering these islands also attracts large flocks of plovers. Here, as on so many of the other uninhabited oceanic islands of the Hebrides, decaying basalts, where comparatively easily burrowed, provide nesting-sites for Manx shearwaters, many of which hatch out their young in the burrows they excavate in some steep talus.

Since the species of seabirds found at the Treshnish may be seen so much more numerously and dramatically on the larger and loftier of the Hebrides, this group has interested me geologically, rather than ornithologically. I remember hearing in the geological classroom some mention of its pre-glacial beaches, found at 90 to 120 feet above present-day sea-level, as on several other isles of the Inner Hebrides—on Mull, for example, on Islay and on Colonsay—beaches loosely so called in this instance because they are entirely devoid of true glacial deposits. Such a beach is well represented on the Bac Mor, that unique member of the group popularly known as the Dutchman's Cap. This type of beach, of course, must not be mistaken for a *patella* beach.

The Dutchman's Cap, resembling at a distance an enormous, broad-brimmed, Netherland hat, is what survives of the cone of a volcano long extinct. Although rising at the Bac Mor's centre from the steep-sided basalt platform surrounding the island to a height of no more than 284 feet, it is the most readily distinguishable of the Treshnish, albeit what remains on Lunga of volcanic outpouring stands nearly 60 feet higher.

Few days in a year permit of an easy landing on the shoreless Dutchman, since the steep cliffs encircling it, though rising to no great height, do not provide even a modified inlet of any kind. For this reason the ruins of human habitation upon it make one wonder how anybody could have survived there for any length of time. Could these be the ruins of summer shielings visited by the Lunga folk in olden times, as has been suggested? What would seem to render this unlikely is the fact that, even on the calmest of days, it would be impossible either to land cows, alive, on this island, or to get them off, alive, at the end of shieling-time. Incidentally, there exists in Mull the tradition that the last human inhabitant of the Bac Mor was a Mull man banished for some serious misdeed, and left to perish there. This is some indication of its general inaccessibility—more difficult to quit with any expectation of survival than to land on.

People with a scientific purpose *do* land on the Dutchman's

Cap occasionally. So also do people actuated solely by adventure.
Even acknowledged mountaineers deem it a worthy objective.
How memorably rewarded was the adventurer who, according
to the appropriate volume of *The Scottish Mountaineering Guide,*
landed there to find in a nest that Challenger golf-ball upon which
a seabird of one kind or another had sat long and unremuneratedly!

With the flora of the Treshnish I am hardly competent to deal.
I might mention, however, that I was pleasantly surprised to find
on their foreshores the oyster-plant (*Mertensia maritima*), its leaves
to the tongue are so revoltingly fishy! As the number of habitats
of this already scarce succulent is steadily decreasing, one would
be unwise to say more precisely where it is to be found.

Among the more interesting of Lunga's mammals are its house
mice, described in 1947 by F. Fraser Darling, the noted naturalist.[1]
James Fisher saw no mice there when he explored this island
ten years previously. Fraser Darling and his observant and
thoroughly competent wife erected their bell-tent at Lunga on a
ledge about 80 feet above the sea, and little more than a good
stone's throw from those ruined, drystone, flat-gabled, roofless
homesteads, which had not been occupied since 1857. 'Within
a few nights of our arrival,' Fraser Darling writes, 'mice of some
sort were coming into the tents and tackling the stores; and it was
not long before they were showing all the cheekiness of the house
mouse and indulging in games. The house mouse is a very playful
animal. Trapping was essential as the stores were not rodent-proof,
and in four months we caught 75 individuals. From the incidence
of this trapping it was obvious that there were always more
coming in, so the total population present must have been
considerable. These mice showed no different features from type
specimens, and it was interesting to note that after the years of
living as field mice, this island race was ready immediately to take
up the traditional existence of house mice again. . . . The Treshnish experience showed that the habits of the house mouse are
strongly hereditary and not greatly altered by education!'

Naturalists, doubtless, are now studying in this context St.
Kilda's house mouse, left pretty much to his own devices since
1930, when I myself, lending a hand with the human evacuation

[1] *Natural History in the Highlands & Islands,* published by Collins
in 1947.

of this island in the autumn of that year, unintentionally sent him a-scuttling.

Apart from Lunga and the Dutchman, the Treshnish Isles are flat-topped. Their modest, elevated plateaux, each resting on a lava platform, are ringed round by perpendicular, or almost perpendicular, cliffs of basalt largely amorphous. On rocky isles like these, such platforms, in so far as they are the breeding-grounds of the Atlantic grey seal (*Halichoerus grypus*), provide the naturalist with the opportunity of studying these creatures without his having to sail, from normal human habitation, long distances to find them. These seals bring forth their young during the none too hospitable months of October and November. How they manage to get ashore in any numbers at the Dutchman seems to me remarkable. But they do breed on it; and, although protected, certainly so far as the proprietor of the Treshnish Isles is concerned, they sometimes are slaughtered there by clandestine intruders, to be carried away by motor-launch. Some years ago Fraser Darling observed such an operation through his glass, but at a distance too great to allow of his reaching the scene effectively.

These seals, of course, are not so numerous at the Treshnish as at North Rona and Sùla Sgeir, which may be regarded as their principal breeding-places. Fraser Darling estimated that the 5,000 seals he counted ashore at North Rona in the late autumn and early winter of 1938 represented roughly half the world's population of the species.

High up on the northern cliff-edge of the Carn Burg Mor, though an islet of negligible size, and with a maximum altitude of no more than 112 feet, are the ruins of a stronghold believed to date back to the days of the roving, rieving Northmen. Tradition has it that its custodian defended it by rolling down from it, upon any band of intruders, huge boulders collected on the shore for this purpose. How such boulders were assembled there in the first instance, and how they provided the defenders with any security, are matters for the student of old-time strategy, who, having landed elsewhere on the island, and comparatively easily, can reach this ruined site from *above*, rather than from below!

Nevertheless, this island is recorded as having been fortified from earliest times. 'Out at sea,' wrote the chronicler, John

Fordun, about 1380, 'at a distance of four miles from Mull, is Carnebrog, an exceeding strong castle.' On the adjacent Carn Burg Beag stands the ruin of a smaller fort with crenellated walls. Donald B. MacCulloch, who knows these waters very intimately indeed, reminds one that both were garrisoned at the time of the Jacobite Rebellion of 1715.

Little of the history of such strongholds is extant, though traditions of that on the Carn Burg Mor are numerous. Seton Gordon, in his *Highways and Byways in the West Highlands*,[1] recalls an indenture, dated 1354, between the Lord of the Isles and MacDougall, Lord of Lorne, under the terms of which the latter surrendered to the former some part of Mull and its lesser islands, but stipulating that the castle of Carnburg was not to be handed over to any member of the Clan Fingon—of the Clan MacKinnon.

Roughly a century and a half later, when the Lordship of the Isles was forfeited, Hector MacLean of Duart was keeper both of this castle and of that on Scarba. In 1504 James IV of Scotland, regarding Carnburg as of some strategical importance, sent a fleet north from Dumbarton to capture it. Setting out with the necessary equipment, which included 'gunstanes', it succeeded in doing so, and in establishing its own garrison there. Nine years later, however, the 'royal castle of Carnburg' fell to Lachlan MacLean of Duart.

That the ancient fortress on the Carn Burg Mor, as well as that on the Carn Burg Beag, were still manned at the close of the 17th century, we learn from Martin Martin (1703). 'They are naturally very strong,' he writes, 'faced all round with a rock, having a narrow entry, and a violent current of a tide on each side, so that they are almost impregnable. A very few men are able to defend these two forts against a thousand. There is a small garrison of the standing forces in them at present.'

The naturalists, J. A. Harvie-Brown and T. E. Buckley, writing so acceptably about the Treshnish Isles two centuries later than Martin, would appear to have got these little forts into their correct historical perspective in a sentence: 'On both of the Cairnbueg islands the remains of the masonry are to be seen, assisting the natural cliff in rendering them certainly most difficult

[1] Macmillan, 1935.

THE TRESHNISH ISLES 113

of assault by a storming party, and they date back only as far as the days of Cromwell, whose forces were for some time successfully resisted by the M'Leans.'[1]

One of the more persistent of West Highland traditions relates to the removal from Iona to Carn Burg Mor of the monks' library. There it is said to have been buried, but whether to preserve it from the rieving Northmen, who so frequently raided Iona, or from the over-zealous Protestants at the time of the Reformation, several centuries later, is not clear. According to Francis Groome's *Ordnance Gazetteer* of 1885, the books and records from Iona deposited there at the time of the suppression of the monasteries were destroyed when a Cromwellian detachment laid siege to it. In all probability, such as remains of the outer, embrasured wall of this fortalice, where it skirts the cliff's edge, was mounted with ordnance. Anyhow, excavations carried out on the Carn Burg Mor in the summer of 1955 by students from St. Andrews University revealed on this isle scarcely a a square yard of soil more than a few inches deep! 'Solid rock,' writes Lucy Menzies in this connection in her volume, *Saint Columba of Iona,* 'is an unlikely repository for the Iona library.'

If this tradition be among the Treshnish Isles' more persistent, that commemorated in the Mull place-name, Arinasliseag, must be among the grimmest. This name signifies the Shieling of the Slicing—of the Slashing. It appears at the north-east end of Beinn Fhada, close to the Lussa River. Thence, after a sharp change in its direction, it finally flows south-east a couple of miles to enter Loch Spelve.

The story goes that MacLean of Duart, at enmity with MacLaine of Lochbuie, swore to ensure that he should have no heir. So he deported him to the Carn Burg Mor, placing with him on that remote isle an elderly woman to look after him. When, several months later, some of Duart's men returned from a visit to the isle, they reported that not only was the prisoner in good health, but also that the woman was heavy with child. This news so disturbed Duart that he straightway had the woman brought back to Mull, and kept under the strictest supervision of a midwife. If the woman gave birth to a male child, he must be killed

[1] *A Vertebrate Fauna of Argyll and the Inner Hebrides,* published in Edinburgh by David Douglas in 1892.

instantly. In due course she was delivered of a daughter. As soon afterwards as was possible, the midwife hurried with this news to her nearest neighbour, in order that Duart might receive it without delay.

Returning to the shieling an hour or two later, she learnt that, during her absence, there had been born to the woman a second child—a boy! In great secrecy this child was carried off to a remote cave in Beinn Fhada, to be placed in the protection of a man and his three sons, then occupying the shieling mentioned. When rumour reached Duart that this heir had been born to Lochbuie, he had the shieling searched, but in vain. Unable to obtain from the father an admission of any kind, the searchers hung about until his sons returned, slaying each of these in turn because of his refusal to disclose the child's whereabouts. After the slaying of the third, they dealt likewise with the father, but not before he had intimated to them that he would die thankfully because none of his sons had betrayed his trust. And so to this day the spot where this gruesome incident took place is identified as the Shieling of the Slashing.

Meanwhile, MacLaine's infant heir was carried off to secret safety elsewhere. When he came to man's estate, his mother told him who he was, and urged him not to divulge this to anybody on Mull. So he quitted that island for Ulster, whence he returned a year later with twelve warriors. Together they came surreptitiously to Loch Buie and took Moy Castle, the MacLaines' hereditary keep. And thus the MacLaines were restored to their ancestral home.

Not since 1824 has any of the Treshnish Isles been inhabited in the accepted sense. In 1817, Donald Campbell and his family, evicted from a croft at Caliach, in the north-west of Mull, settled on Lunga and remained there until 1824. Three of Donald's children were born on Lunga. One day in 1824 Donald, afloat in his little boat off the Dutchman's Cap, shooting geese for the family pot, was caught in a squall which bore him away to Coll. When he did not return to Lunga after an absence of four days and four nights, the anxiety his wife and children were sharing was turned to despair. They had given him up as gone forever. His wife, in her demented state, had allowed the fire to go out. She had no means of re-lighting it. When Donald did return home, he and his dependants agreed that never again could they risk a repetition of such an ordeal, and that therefore they must quit.

The Treshnish Isles have been uninhabited ever since, except for seasonal residents. A well-known Mull fishing family of the name of Robertson (still very much alive in Tobermory, though no longer fishing) used to spend the summers on Fladda while lobster-fishing in Treshnish waters. Each season it despatched to Billingsgate an average of 4,000 lobsters.

Today, Lady Jean Rankin, widow of Niall Rankin, who purchased the Treshnish Isles some years ago and put Shetland sheep on them, lets their grazing to a first-rate Dervaig family named MacLean. With a stout fishing-boat of the half-size, ring-net type, the MacLeans are doing well among the Treshnish with scampi and lobsters.

CHAPTER EIGHT

The Loch Lomond Nature Reserve

LOCH LOMOND, world-renowned for its singular beauty, for its historic, romantic, and literary associations, is by far the largest freshwater sheet in Great Britain. With a surface area of 27 square miles, it is 5½ square miles larger than Loch Ness, its runner-up, although only a fifth the size of Lough Neagh, largest lake in the British Isles.

Shared by the shires of Dunbarton and Stirling, Loch Lomond's sailing length from the pier at Balloch to that at Ardlui is 24 miles. At its southern, triangular end it attains, from east to west, a maximum width of 5 miles. For a dozen miles or so, the northern half of its length, where it laves the foothills of territory so intimately associated with the exploits of my freebooting fore-bear, Rob Roy MacGregor, has an average breadth of not much more than a mile. At the Pulpit Rock, a couple of miles south of Ardlui, it contracts to a few hundred yards.

The loch's height above sea-level is 25 feet. Its greatest depth is 623. Except for the relatively flat territory bounding it on the south-east to form part of the historic parish of Buchanan, it is enclosed by lofty mountain-ranges culminating on the one hand in Ben Vorlich and Ben Vane, and on the other in Ben Lomond, one of the noblest of Highland peaks. Fed plenteously at its northern end by the impetuous River Falloch, by the slow and sinuous Endrick Water at the south-east, and by the numerous mountain-streams and tumbling cataracts descending precipi-tously to it along most of its perimeter, its overflow, issuing from it at Balloch, is the River Leven, which runs seven miles through the industrial Vale of Leven to meet the Clyde at Dumbarton. Fed *plenteously*, I have said. 'Weather is always tricky at Loch Lomond,' remarked a Nature Conservancy officer when I drew his attention to the lowering clouds one midsummer's day. 'With a 70 inch rainfall, there's bound to be an awful lot of days unsuitable for photography.'

*　　　*　　　*

Of Loch Lomond's thirty islands, islets, and merest eyots, six of the smallest, albeit devoid neither of historical nor of archaeological interest, lie north of Ross Point, the wooded promontory reaching well into the loch roughly eight miles north of the pier at Balloch. The others, ten of them considerably larger and correspondingly more important, are distributed about the loch's southern end, upon that triangular expanse already referred to. In that expanse, which broadens gradually toward the south, are five islands which, together with a comparatively low mainland area of 436 acres, fringed by the south bank of the winding Endrick Water but a mile to the east of them, comprise the Loch Lomond National Nature Reserve, part of which, as in the case of Loch Lomond itself, lies in Dunbartonshire, and part in Stirlingshire. The islands are Inch Cailleach (138 acres), Torrinch (18 acres), Creinch (15 acres), Clairinch (15 acres), and the crescent-shaped acre known as Aber Isle. In 1962 the first mentioned was purchased by the Nature Conservancy. Two years later, Clairinch came appropriately into the possession of a Buchanan who as appropriately presented it to the Buchanan Society, since the Clan Buchanan, centuries ago, adopted this island's name as its warcry. Clairinch, along with Inch Cailleach and some other Loch Lomond islands with which we are not concerned at the moment, forms part of the historic parish of Buchanan. The other islands included in the Reserve are the property of Sir George Leith-Buchanan.

These islands are reached usually and most conveniently from Balmaha, a spot as famed for its Pass in the heyday of the cateran Clan Gregor as it is today for MacFarlane's Boatyard and the numerous jetties adjacent thereto. Off these jetties are moored scores of sailing-craft of every shape and size. Here, at Balmaha, one finds a forest of shining masts and fluttering pennants, of canvases in every colour, each contributing in its own intimate way toward rendering Balmaha, on a day of sunlit cloud, one of the most refulgent of scenes. Indeed, during the summer and autumn

months, boating and yachting, now so popular a recreation on
Loch Lomond, lend to Balmaha a gaiety rivalling that of any of
the traditional seaside resorts on the Firth of Clyde. Red sails in
the sunset are an integral part of the scene.

*   *   *

Having regard to these islands' limited area, they offer a
remarkable diversity of interest. South-westward across Loch
Lomond, and in alignment with Conic Hill, which lies behind
Balmaha, there runs that mighty dislocation known to geologists as
the Highland Boundary Fault, evidence of which is so clear on
Inch Cailleach where, as it were, it tends to separate the north-west
third of that island from the other two-thirds. In so far as this
well-defined geological line marks the position of what must be
regarded as the greatest discernible dislocation in Britain, it also
demarcates two entirely different types of scenery. To the south-
east of it, where the rocks have been set sharply on edge against
the flanks of those on its north-west, lie the comparatively low
hills of Old Red Sandstone, of kindred conglomerates, and of
volcanic rocks, all interspersed with those rich plains characterising
much of this lowland countryside. To the north-west of this
dislocation, on the other hand, stretches a chaos of crumpled
gneisses and schists. A sea of mountains, as Archibald Geikie
describes it—mountains rolling away to Cape Wrath in wave after
wave of gneiss, schists, quartzite, and other crystalline masses.[1]

Here and there on Inch Cailleach, particularly at its few relatively treeless spots, although also in the dense undergrowth
associated with them, one sees the clearest evidence of the Highland Boundary Fault in the gritty slabs or slices of rock resembling gigantic puddings with huge, chocolate-coloured raisins irre-
gularly distributed throughout them. Indeed, one finds on this
island splendid specimens of the Reserve's main geological
features. Between Tom nan Nigheanan, its dominating ridge,
which overlooks Clairinch and the Kitchen, but a few hundred
yards to the south-east—between Tom nan Nigheanan and the
parallel Church Ridge runs, in a north-easterly and south-

---

[1] *The Scenery of Scotland*, first published by Macmillan in 1865, and
again in 1887. No Scotsman's library is complete that does not include
a copy of this superb piece of scholarship.

westerly direction, its central valley. This extends roughly from
North Bay, at one end of it, to the sandy Port Ban beach at the
valley's other end, sheltered by what are now designated the
South and West Promontories. The distance between these
extremities, linked by a good path mildly interrupted about halfway by an outcropping crest, is not much more than half a mile.
On each side of the Port Ban the outcropping and steeply tilted
Lower Red Sandstone conglomerates are seen to advantage. In the
little crag of dolomite serpentine protruding from the shore a
hundred yards to the north of West Promontory may be seen at a
glance a relatively small, but nonetheless interesting, feature of the
Highland Boundary Fault.

These islands' chief scientific interest, apart from their geology,
resides in their woodlands, which are composed mainly of oaks,
with a healthy sprinkling of Scots pines. Tree-felling for housing
purposes had already begun on Inch Cailleach just prior to the
Conservancy's purchasing it. Thus the island was rescued from a
denudation that otherwise would have been inevitable.

In springtime the undergrowth of these woodlands, almost
wholly ungrazed, is an enchanting sea of wild hyacinths. In summer the brackens take over in a masterly way. Festooned with dog-
roses and honeysuckles, they stand seven or eight feet high, out-
topped only by foxgloves aspiring from deeply shaded places to
reach what sunlight they can.

The paths beneath one's feet on Inch Cailleach, sometimes
thick with mud in places after heavy rain, often show deer-slots
recently made. Where the brackens are particularly dense, these
paths afford deer the easiest passages across it. That deer, though
so seldom *seen* on any of these islands, make daily use of at least
some of these paths was obvious to me when, on the second
of two successive days' visits made a year or two ago, I noted
fresh slots in a dry path upon which, the previous day, showed
no more than my own footprints.

The islands' present stock of fallow deer is descended, doubtless, from those members of the species introduced and re-
introduced to some of them at varying intervals since the 16th
century. The earliest recorded introduction would appear to be
that of 1530. Today, as in olden times, the deer move freely to
and from and between these islands, swimming usually at first

light. For much of the year the islands appear to carry does only; while from time to time a herd of bucks only is seen on the Reserve's mainland territory. From late summer onwards, the bucks return to the islands for the October rut, though in 1967 a fine buck was spotted on Inch Cailleach on July, 5th. This, I believe, is the earliest recorded arrival.

Just as thickly encumbered are such paths as exist on the Reserve's other and less frequented islands. On Clairinch, for example, where a landing at either of its tiny, adjacent beaches is easy, the paths leading inland from them are difficult to locate on account of dense vegetation. The hazels on Clairinch reach down to, and largely overhang, the shore-line.

It is from these little beaches that one sees so plainly, and at no distance, the almost perfectly circular, 80 feet in diameter, islet known as the Kitchen. This crannog of more than two millennia ago is thought to have received its present name from the low, fire-blackened stones arranged at its centre by picnickers in olden days, and still used by their picnicking successors.

*   *   *

Directing one's attention in a south-easterly airt from Clairinch and the Kitchen, one spots, at a distance of a mile or so, Aber Isle. It probably takes its name from that vanished Loch Lomond-side village, the former existence of which is commemorated by the inclusion on the Ordnance Survey map of the place-name, Townhead of Aber, situated but a mile and a half to the south, within a few hundred yards of Ross Priory. This isle is by far the smallest of the Loch Lomond Nature Reserve's components. Furthermore, it is the only one upon which a landing is at all times tricky, even from the smallest boat, and certainly from a *motor*-boat, since it lies surrounded by rocks and reefs deceptively submerged except when the loch's level is low enough to expose at least the highest of them. Powerful cabin boats, because of their speed and greater draught and the vulnerability of their pro-pellors, must be handled dexterously and with foresight in this locality, especially when there is the slightest wind. The briefest invigilance off Aber Isle would spell disaster.

*   *   *

Loch Lomond and its several islands have long engaged the

TRESHNISH ISLES. The ruins on Lunga

The Dutchman (Bac Mor) from Lunga. Shags and Puffins are seen
sunning themselves in the foreground. Lunga is now used solely for
grazing. Its last inhabitants left in 1824

THE TRESHNISH ISLES. The Carn Burg Beag from the Carn Burg Mor on a hazy, Hebridean day. Mull is seen dimly in the background to the left

THE TRESHNISH ISLES. Ruins on Carn Burg Mor. 'Out at sea, at a distance of four miles from Mull, is Carnebrog, an exceeding strong castle'

interest of ornithologists, partly on account of the rare migrants to be seen among them from time to time, and partly because of their wildfowl. The latter are still sufficiently numerous to attract poachers in numbers necessitating the wardens' unremitting watchfulness. Up to date, no fewer than 147 kinds of birds have been recorded on the Reserve, some of them extremely uncommon. Buzzards nest regularly on its islands. Capercaillies do so occasionally.

The corner of the Reserve which Glasgow's bird-watching enthusiasts find most attractive is, of course, the sand-bar at the Endrick's mouth. Close at hand is Tern Island, to which the Conservancy might well give some attention if the common tern is not to be lost to the loch as a breeding species. Although at one time firmly established on some of its islands, today, owing to the increasing encroachment of picnickers, it is losing the battle for its nesting habitat of sand and short turf.

Loch Lomond's late-summer drop of about 25 feet O.D. synchronises with the autumn migration of waders from the North, emphasising the ornithological importance of an area to which Sir George Leith-Buchanan first drew attention about the middle of the 19th century. In looking through this proprietor's records, however, one sees how he blasted every creature that moved between his doorstep at Ross Priory and the Endrick's mouth. His bag included such rarities as Bonaparte's Gull and the North American Pectoral Sandpiper.

In recent years the Endrick locality has become increasingly the haunt of wintering geese, especially of grey-lags, and to a lesser extent of pink-feet and white-fronts. The grey-lags travel between the Endrick's marshlands and the low, flat tract of Stirlingshire known as the Flanders Moss. Since Robert Gray published in 1871 *The Birds of the West of Scotland*, there has been an enormous increase in the number of grey-lags wintering here. Formerly noted as an occasional visitor during particularly hard winters, its wintering numbers now range from 600 to 800. Small wonder local wildfowlers showed their resentment when the Nature Conservancy's NO SHOOTING notices went up! Now they have come to recognise the grey-lags' territory as a wildfowl refuge; and 'incidents' with the wardens are rare.

Ducks also frequent this region, nesting there in considerable numbers, some of them precariously because the height of the meandering Endrick is controlled by the level at which Loch Lomond itself, at its natural outlet, is maintained. If heavy rains swell the streams and cataracts entering it, thus suddenly raising its surface to a height at which its flood-waters can escape at the overspill to augment at Balloch the River Leven, much of that low, nesting territory of bushes and stunted trees becomes inundated. Consequently, eggs already laid are apt to be swept off to the loch itself. On the excessive flood-waters that overflowed the lower Endrick's banks in the spring of 1961, scores of wildfowls' and of waders' eggs were floated lochwards.

Incidentally, it may not be known generally that the winds descending from the mountains often whip Loch Lomond up into a turmoil in an impulsive moment, as they are apt to do at all seasons of the year, and with little or no warning. The huge trunks and roots of trees beached high above the loch's normal height testify amply to this, both on its islands' shores, and on those of the Conservancy's mainland territory, especially when the winding course of the Endrick Water allows of their being washed some way inland, upstream.

*   *   *

No less diverting than the Reserve's ornithology is its botany. Its list of native plant species, compiled as recently as April, 1967, by the naturalist, E. T. Idle, its principal warden, in cooperation with a number of enthusiastic botanists is indeed formidable. Up to date it comprises 308 vascular plants, 96 mosses, 45 liverworts, and 67 lichens.

*   *   *

At one time or another, and for a variety of purposes, several of Loch Lomond's islands have been inhabited, and over a great number of years. Remains of ancient structures are to be found on two of those belonging to the Reserve—on Clairinch and on Inch Cailleach. Those on the former are small and of minor significance. Those on the latter, however, are substantial and of considerable archaeological account. They consist of the foundations of an ancient and ruinous church, and of the still more ancient burial-ground situated immediately to the south-west of it.

Locally, both church and burial-ground are held in the highest esteem. With their preservation and care the Nature Conservancy is conscientiously concerned.

That its scheme for conservation and improvement is proceeding satisfactorily is seen most clearly, perhaps, on the mainland portion of this Reserve. There the broken-down gate and rusty barbed-wire fence one knew have been replaced with an attractive wooden paling and a five-barred gate, and also with a small wicket affording easy public access. The work of fencing off cattle from the under-grazed woodland is already in hand, as is a programme of planting and thinning young trees. The present owner of this territory, with whom the Conservancy has a Nature Reserve Agreement, did not realise until quite recently that the alluvium of the Ring Point, where it reaches down to the Endrick's mouth, is capable of sustaining a rich grazing. The systematic grazing of this area, by checking the spread of scrub and of the coarser grasses, will enhance its importance as a habitat for open-nesting birds.

*   *   *

No islands in the care of the Nature Conservancy are as accessible as are these; and there exist no restrictions about landing on them. But accessibility can have its drawbacks, particularly from an amenities point of view. It has certainly left these islands rather much at the mercy of vandals and litterers. Of course, no reasonable person would expect completely unrestricted access to a Nature Reserve, since such would speedily result in its ceasing to be a Reserve, certainly in the sense which Conservancy implies. In regard to these Loch Lomond islands, all that is required of the visiting public is that measure of decorum nowadays so disconcertingly lacking. It is lamentable that, quite irrespective of class, several decades of compulsory and costly education should have inculcated so little self-discipline in matters of tidiness, orderliness, and hygiene. During the absence from Loch Lomondside of either of the Reserve's wardens, for even a day or two, the amount of litter in the shape of tins, bottles, canisters, cartons, metal tops, and even sanitary towels, is disheartening. It often takes a returning warden an hour or more to clean up on Inch Cailleach alone what unrestricted democracy has thrown down on the shore by the landing-stage there in a single afternoon, much of it within a few yards of the baskets provided for it. I am not unaware, of course, that some of the litter at such spots can be attributed to seagulls raiding these baskets with as little sense of tidiness as have most human beings let loose on their own public properties.

A firsthand account of public misconduct on these islands was given me by Mr. John Mitchell, one of their wardens, when, by motor-boat and on foot, he conducted me on a tour of the properties concerned. 'It's here we meet Joe Public face to face,' he remarked as we stepped ashore on Inch Cailleach from his motor-boat, *St. Kentigern*. 'Last Monday, despite all our informative signs and suitably placed litter-baskets, I picked up off this pocket-sized Port Ban beach two sackfuls of rubbish dropped the previous sunny day, immediately following National Litter Week!' I regarded such observations of his as sufficiently serious to warrant my jotting down anything else he might have had to say on the subject. 'It's discouraging to have a chap find that he can be too efficient in cleaning up after the litter louts. Loch Lomond's "floating population" now regards Inch Cailleach as some sort of God-sent or corporation-sponsored Refuse Disposal Centre. Launches often pull in for the sole purpose of dumping old junk, and are off immediately. In addition to ginger-beer bottles and tattie-crisp bags, I now have to deal with rolls of worn lino and mounds of moth-eaten curtains. There hasn't been a discarded bedstead yet; but this is only a matter of time. Thank goodness for "the moat" separating us from the mainland at Balmaha, where an increasing army of Glaswegians stands poised every weekend, like a Normandy Invasion Force.'

*   *   *

I conclude this chapter with a few words on the archaeological significance of Inch Cailleach, Isle of the Old (or Cowled) Women, which dates from the early years of the 8th century, when the widowed Kentigerna, a missionary from Ireland, retired there and, in 733, died there. Five centuries were to elapse ere the foundations of the church dedicated to her were laid at the heart of that island, deeply sequestered among its hoary oaks, where already lay an ancient place of burial. These are the foundations which the Islay Archaeological Survey Group has been engaged in excavating. By great good fortune, the Conservancy has had the services of this voluntary band of archaeologists who, in recent years, have been removing a centuries-old growth of vegetation to discover precisely what lies hidden beneath it. While so doing, they have unearthed some interesting relics, among them an early 14th-century silver penny of English origin. As a result of their first two field seasons, the Ministry of Public Buildings and Works scheduled this burial-ground as an ancient monument. Some attention has been devoted to the site of the old farm-buildings and of the arable land that once existed on Inch Cailleach. Regrettably, at the time of writing, no final report on any of these excavations has been published.

My own personal interest in this isle derives from the ancient burial-ground already mentioned—or, rather, from the Grey Stane, a particular tombstone in that historic burial-ground, where may be found tombstones dating from the 12th century to our own 20th. These, for the most part, mark the graves of various clansfolk, natives of the surrounding countryside, many of whom were MacGregors.

By the Grey Stane, so memorably alluded to in Sir Walter Scott's *Rob Roy*, members of the proscribed Clan Gregor in olden days took the oath of honour and fealty. 'By the halidom o' him that sleeps beneath the Grey Stane on Inch Cailleach!'

In the Highland home of childhood and boyhood, one was called upon to swear by this ancient MacGregor emblem, rather than upon the Bible.

CHAPTER NINE

Mingulay

EILEAN MO CHRIDHE, Isle of my Heart, the old folks called it. They scarcely knew any other. Mingulay was their world—Mingulay and the encircling seas, sometimes halcyon, oftener tempestuous. After Barra itself, theirs was the largest of the archipelago long known as the Barra Isles. To the south of them lay precipitous Berneray, or Barra Head: to the north lay Pabbay, Sandray, Vatersay, and, of course, Barra.

Whereas Mingulay's menfolk had reason to visit these other isles from time to time, usually when fishing off them or when attending to livestock pastured on them, an occasional trip to Castlebay in the summertime was the farthest its womenfolk wished to get from what they had been accustomed to, and content with, all their lives. Even then, in 1911, when the last members of this isolated community were transferred to Vatersay, the little metropolis of Castlebay, a mile or two farther north, was still largely a place of black-houses in no way different from those comprising their own township. In any case, the journey to Castlebay, accomplished of necessity at that time in an open fishing-boat under sail or oar, was a long and tedious one of fifteen miles, and in seas not infrequently surging, even in summer. Thus Mingulay's womenfolk were loth to leave their own intimate, pastoral simplicity, even for a day. Isle of their Hearts provided all they asked of life. When eventually they had to quit, it was as though the very roots of their existence had been torn up. They were being wrenched from everything that, hitherto, had nourished, if but meagrely, both mind and body.

What an isle was theirs in the Elysian days of the short-lived summer! In those sheltered hollows, to which the tethered kine were not granted access until the last hay-crop had been harvested in the late autumn, the wild flowers of its pastures then stood knee-deep. Mingulay then, to its simple inhabitants, was, indeed, Arcadia. Never had Pan piped in country more delectable. Here the shepherd's god could have found rustic contentment not a whit inferior to that of his Peloponnesian domains. Here resided happiness supreme, despite the exiguity of their island's economy. What the eye had not seen, the soul had not pined for.

The world beyond this isle and its immediate neighbours was one of far horizons—Coll and Tiree, low-lying to the south-east; Ben Hiant, in Ardnamurchan; Ben More, in Mull; the Coolins of Rum and the Coolins of Skye, usually discernible when donning their nightcaps; lone St. Kilda, in the clear weather prognosticating rain, inhabited by people whose lot differed little from their own.

Of course, the islanders may not have known too precisely which of these distant landmarks was which. That territories so named existed, they never doubted, however, since they turned up in the legendary and folklore recounted round their own peat-fires during the winter evenings. It was left to visiting geologists, naturalists, and storm-stayed strangers to tell them just what, under favourable atmospheric conditions, might be seen in this direction or in that. But what cared *they* for accurate identification in this regard? All that really concerned them was their own *Eilean mo Chridhe*.

The isle of which I write is Mingulay. And how sweetly sounds its name! Yet, one cannot reproduce it accurately in English phonetical equivalents. The accent falls on the first of its three syllables, which is pronounced like two-thirds of 'mingle', except that the vowel-sound lies somewhere between *i* and *e*. In other words, Mingulay belongs to the realm of phonetics none but the Gaelic tongue ever quite achieves. That, perhaps, is why the merest mention of it is music to the *soul*, as well as to the ear.

For those who knew Mingulay in the early years of the present century and are still alive, it connotes that vanished era when, in the brief season of smooth seas and flowery pastures, life throughout so much of a world comparatively recently modernised, mechanised, secularised, and bedevilled was idyllic in the truly Greek sense. It also connotes the heart-beat reduced in tempo almost to that of the hibernating mammal, since, in time of winter

wind and wave, the day had scarcely dawned ere it fell a-dusk.
What could the natives of an isle thus situated do, then, after they
had attended to the primary and primitive needs of man and
beast? Little, except remain drowsily indoors. For weeks on
end, if not actually for months, gales and rains and pounding seas
kept them there, a-dreaming over their peat-fires.

*   *   *

Mingulay lies but half a mile to the north of Berneray, most
southerly of the Outer Hebrides. After Barra and Vatersay, it is
the third largest of the group of major islands listed at the
beginning of this chapter, and known collectively as the Barra
Isles. At the close of the 16th century, according to Dean Monro,
it pertained to the Bishop of the Isles, as did its neighbours. At
the time of Martin Martin's celebrated peregrination of the
Hebrides (*circa* 1695), *all* the islands south of Barra were termed
the Bishop's Isles. To what extent they were the Bishop's, rather
than The MacNeil's, it would be difficult to say. Certainly in
subsequent centuries they belonged very emphatically to the
MacNeil Chiefs of Barra, whose clansfolk inhabited them almost
exclusively. Today, of course, Mingulay, like the rest of the
Barra Isles, forms part of the County of Inverness.

Now for a few measurements such as may help the reader to
visualise not only Mingulay's location, but also its geographic proportions. It lies roughly 90 miles north of Ulster, and the same
distance west of Oban, in Argyllshire. Its area, according to our
painstaking Ordnance Surveyors, is 1,508 acres, in addition to
which it possesses, in its beautiful bay of sheen-white sand, some
48 acres of foreshore. Its maximum length and breadth are,
respectively, $2\frac{3}{4}$ and $1\frac{1}{2}$ miles. Its coastline measures $12\frac{1}{2}$. Except
for the relatively small part of it represented in the bay I have
mentioned, it is so remarkably cliffy, lofty, precipitous, and inaccessible as to be the breeding-place of countless seafowl. In
Britain, only the sea-cliffs of Hirta, of the Boreray of St. Kilda,
and of Foula excel those of Mingulay as regards altitude, magnitude, and thronging bird-life. Not greatly inferior are those of
Berneray, its neighbour to the south. Berneray, with a precipitous
coastline of nearly six miles, has an area of 446 acres. That is
to say, less than a third of Mingulay's. Pabbay, the isle im-

mediately to the north of Mingulay, is larger than Berneray by 144 acres.

In splendour, as also in the prodigality of the seabirds inhabiting them at nesting-time, stretches of Berneray's cliffs vie with those of Mingulay. On the south and west of the latter island is an area of between 700 and 800 acres of sheer rock-face. The island's highest point, situated near its western extremity, is 631 feet. Close to it, and on the brink of its astonishing cliffs, is Barra Head lighthouse, the lantern-tower of which stands at an altitude of no fewer than 683 feet above highwater mark. Situated at the southernmost tip of the Outer Hebrides, its white, occulting light, though with a candlepower of but 39,000, illumines for 60 seconds, every 60 seconds, a wide expanse of ocean. In fact, its great height renders it visible on the clearest of nights at roughly 38 miles. Today, Berneray's three lightkeepers are its sole human population.

A tremendous chasm at no distance from the lighthouse, 300 feet in length and nearly twice as many across, terminates in a cave, the dimensions of which are not accurately known. Perhaps, some spelaeologist, after reading this, may feel prompted to fill in accurately this gap in human knowledge. If I weren't too fully occupied with kindred matters in other parts of the world, I'd see to this myself!

Seas surging beneath Barra Head lighthouse have been known to deposit fish on the bare rocks at its base. That is to say, at a height of more than 600 feet! I myself have witnessed here, at Barra Head, infinitely mightier seas than at the lighthouse perched on the Butt of Lewis, the other extremity of the Outer Hebrides.

The table below enables one to see at a glance a few of the more important dimensions of the five islands known in olden times as the Bishop's Isles. These may be of interest to the reader to whom geographic detail is important:

	Land	*Area in Acres* Inland Water	Foreshore	*Length of Coastline*
BERNERAY	446.7	Nil	10.78	5¾ miles
MINGULAY	1508.2	Nil	47.7	12½ miles
PABBAY	560.85	Nil	52.3	7½ miles
SANDRAY	1002.9	1.9	53.5	9¼ miles
VATERSAY	2272.5	.8	230.3	19½ miles

E

Mingulay has three appreciable heights. Carnan, highest of them, situated on the west side, attains an altitude of 891 feet. Inland, and at no great distance to the south-east of Carnan, Hecla rises 700 feet. The northern part of the island reaches in MacPhee's Hill a height of 735 feet. From the cliffs' edge at the western side, then, at nearly 900 feet, Mingulay slopes down to sea-level on its east side, and at the bay of white sand already alluded to—at Mingulay Bay, as it is called. Another slope, but of lesser extent, falls to the south, in the direction of Berneray.

These slopes and their respective directions ensure that the island's sheep-stock can find adequate shelter from almost any wind that blows—an important matter on territory so windswept, and with a soil so little conducive to growth, that neither tree nor shrub can obtain a footing. In addition to such natural cover as the sheep may avail themselves of, they are free to resort to the shelter afforded by the walls and stony hollows of the island's ruined village, where such of it has not yet been entirely over-whelmed by drifting sand.

Here one might say a word on how important a bearing on the breeding of sheep is the direction in which slope their hill pastures. Since those on Berneray slope toward the north, they are a couple of weeks later in attaining the succulence which Mingu-lay's herbage already has, owing to the difference in the general direction of their inclination. The rams are placed on Mingulay (20 to 24 rams to roughly 700 ewes) with a view to ensuring the fall of lambs early in April, when the grass is just coming to its best. Berneray, with its excellent herbage, carries from 300 to 350 ewes. The rams are not introduced to that island, however, until roughly two weeks after they have been liberated on Mingulay, for otherwise the lambs would be born before the young, fresh grass has grown sufficiently to maintain the mothers adequately in milk, and this would result in casualties among the crop of lambs. Nevertheless, the sheep and their lambs are handled earlier on Berneray than on Mingulay. This is due to three advantages Berneray possesses over Mingulay. One of these is its pier, which renders loading and un-loading so much easier. Another is the additional help available on Berneray: the light-keepers there are always happy to lend a hand with the sheep. Thirdly, the gathering of the sheep is so much easier than on

Mingulay, which is thrice its area, and in other ways more difficult to range over.

Of the not inconsiderable bearing of bird-life on lambing where islands like these are concerned, I learnt something a few years ago from a friend, Mrs. Peggy Greer, who, in 1937, purchased Pabbay, Mingulay, and Berneray, the three southernmost of the Barra Isles. Peggy retained them until 1951. I may as well introduce her to you here, since I shall have a certain amount to say later about her activities on Mingulay and about her ultimate disposal of it and of the other two islands included in her purchase.

The hatching season of the seabirds infesting the cliffs of these Hebridean isles is roughly April. Peggy, therefore, timed her lambing on Mingulay to synchronise as far as possible with the birds' hatching of their young along the upper tiers of its stupendous cliffs. These birds, at nesting-time, can be quite aggressive. On the approach of an intruder, be he human or animal, they will create a deafening din, and even peck at the intruder sooner than take to the wing. In this wise, seabirds nesting on the dangerous cliff-tops restrain sheep that otherwise would venture to reach temptatious blades of herbage on rocky ledges whence their following lambs cannot retreat, and whence even many an adult ewe cannot retreat either. Peggy had it on the testimony of her shepherd, however, that a courageous ewe will often defy the noisy threats of sitting seafowl, and descend perilously to tufts of alluring herbage. In order to prevent this, she fenced in the entire length of Mingulay's more precipitous cliffs. This entailed some two miles of fencing—no simple undertaking, especially in those parts where solid rock lies but an inch or two below the sward, thus necessitating the seating of iron stakes in the rock, and the pouring of hot sulphur round the base of each, when erected, in order to set it firmly.

Still, it is doubtful whether the casualties among Peggy's sheep were not due to the ruthlessness of the black-backed gulls rather than to accidents on the cliffs. This particular species of gull is the great pest of sheep, which explains how hornless sheep cannot survive long on isolated places like Mingulay. The gull seizes the sheep's forelock; and the sheep, having no horns with which to butt at it, heaves its head from side to side until, in exhaustion, it falls victim. Its eyes are immediately picked out. Soon afterwards the sheep unwittingly commits suicide by falling over some cliff-edge. Before long, the gull has drawn its intestines out through the anus. On these it feeds, leaving the carcase thereafter, and returning to it when decomposition has set in sufficiently to allow of its tackling it without much difficulty.

When witnessing the black-back's ferocity on a recent occasion, I could not but put to myself the question that William Blake, in contemplating the tiger, posed in this identical context: *Did He Who made the lamb make thee?* Who created such a world of calculated cruelty and carnage? The All-loving, All-merciful, All-powerful God allegedly revealed to the Christians through their Crucified Christ? Whiles, I ha'e ma doots!

*　　　*　　　*

The Atlantic side of Mingulay consists for the most part of a series of mighty cliffs falling perpendicularly and so smoothly that neither bird-life nor plant-life can gain a foothold upon them. This is particularly true of the cliff called Aoineag, or Biulacraig, a black, frowning precipice 753 feet in height. Such seabirds as throng the adjacent stacks, crags, and skerries during the nesting-season are entirely absent from Aoineag for the obvious reason that they cannot alight, far less lay an egg on it. To look up the face of this bastion from a small boat brought close inshore immediately below it, as I myself have done on more than one occasion, is assuredly one of the most sobering of human experiences. Imagine something soaring perpendicularly out of the ocean to a height of, say, Arthur's Seat, and examine your own insignificant self afloat at the base of it, looking up and up! Unless you have had previous acquaintance with such situations, you would probably be overcome by something more terrifying than you hitherto had known in your life.

Of course, there are few days in a year when, without incurring a considerable measure of peril, the encircling waters are calm enough to permit of such an adventure. On the other hand, when it *can* be done with safety, it is well worth doing. One requires for it that degree of calm enabling one to sail a small boat in through the enormous sea-caves near the base of Aoineag, and out through another—a thrilling experience I myself had when, some forty-five years ago, I paid my first visit to Mingulay as a passenger aboard the lighthouse relief-boat then on one of her regular trips between Castlebay and Barra Head. We emerged by a passage separating Gunamul and Arnamul, two huge stacks isolated from the main island. The precise location of this sea-tunnel is indicated on the Ordnance Survey map by the words, Natural Arch. The stacks I have mentioned, like the cliffs adjacent thereto, are thronged with seabirds. In contemplating them a few years ago, I came to the conclusion that they might well present the rocker—the cragsman—with climbs as formidable as those involved in the ascent of Stac Lee and Stac an Armuinn, the celebrated, gannet-haunted stacks at Boreray of St. Kilda, which only the most intrepid cragsmen in the world ever succeed in ascending and descending.

Were Mingulay and Berneray more accessible, there is little doubt that they would attract rock-climbers in search of difficulty and danger. Their cliffs could certainly provide climbs which would test the daring and endurance of the most skilful cragsmen, although one should bear in mind that the natives of these two islands, during their inhabited days, thought little of them, having been bred to traversing these precipitous rock-faces, literally from boyhood, when fowling and collecting seabirds' eggs. Those whose pleasure it is to seek out the most perilous routes up and down cliff-faces would certainly find what they wanted on Mingulay, where even the most agile cliffsman cannot afford to take liberties. Only in the cliffs of Hirta, of Boreray, of Soay, and of Foula does one find in Britain anything of the kind so magnificent and awe-inspiring.

Much more impressive than Gunamul or Arnamul, and also less accessible, is Lianamul, a stack lying close inshore, roughly a mile to the north of them. Harvie-Brown, who visited it in 1871, and again in 1887, described it as the most densely packed guillemot station he had ever beheld. 'This fact,' he wrote, 'is no doubt owing to the unsurpassed suitability, regularity, depth, and number of the breeding ledges, along many of which two men could crawl abreast on hands and knees, with a roof of solid rock above, and a floor of equally solid and horizontal rock beneath. Deeply cut into the cliff-face are these great horizontal and parallel fissures. There is no tilting outward of the strata, no "fault" in their regularity, while the top of the Stack is deep in rich sorrel and seapink-covered mould, the accumulation of many years of bird excreta, and which is tunnelled and honeycombed in every direction by Puffins.' These birds, he goes on to say, had completely ousted from the summit of Lianamul the Manx shearwaters that once occupied it.

At this time a score of sheep grazed the summit of Arnamul, and five that of Lianamul. Formerly, the latter stack had been rendered accessible by means of a rope-bridge suspended between it and the parent island of Mingulay. By the 1880s this means of access had gone. The natives resorting to it thereafter, for one purpose or another, did so by way of a landing on the seaward side, which could be reached only under exceptionally smooth conditions. A perilous climb then confronted them.

These stacks became separated from the parent island, as in the case of all such lofty and isolated structures, by the disintegration, from summit to below sea-level, of the softer rock once intervening. This rock was gradually dissolved by the salt water of the ocean, or carried away and strown elsewhere by the co-operative actions of wind and wave, tide and current. The result of such epigene and marine weathering is clearly seen at Mingulay, where an enormous trap-dyke has been eroded away. There is a splendid example of the same process at the island's south-western extremity. Here a peninsula with precipitous cliffs sheering 400 feet to the sea, linked to the main island by a narrow, ridged isthmus, is slowly being worn down through geological time to sea-level. On the peninsula itself are the ruins of Mingulay's ancient *dun,* or fort. Across the trap-talus of the connecting isthmus, the inhabitants of olden days had erected, as an outer defence, a considerable wall, only a fragment of which survives. To this *dun* they retired for protection against marauding corsair and raiding neighbour. Similar weathering is very apparent in the north cliffs of Berneray, where it has created, and is inexorably enlarging, a deeply receding chasm.

In time of storm, Mingulay's cliffs present a scene of unparalleled splendour. Here the gales blowing in from the unfettered Atlantic bring the seas a-crashing against these gigantic cliffs with such violence that foam and spindrift are sent flying over their tops at 700 or 800 feet. Anybody standing by the cliff-

edge when the ocean is pounding Mingulay's western side would
be soaked to the skin in a moment by flying spume and spray,
even at this height. A visitor to the island in 1911 recorded that
he had actually gathered on the sward, at the top of the cliffs,
handfuls of tiny fish thrown up this great height during a
westerly gale.

Of the several sea-caves penetrating Mingulay's western cliffs,
the natives harboured a superstitious dread, believing them to be
the abodes of ghosts and goblins, of the water-horse and kindred
monsters of the deep. Only the bravest among them—their cragsmen and lobster-fishermen—ever set eyes on these caves; and
few, *even of them*, dared enter such subterranean places. One
cannot suppose that any woman on Mingulay ever found herself
in a boat below its western cliffs, where lie these dark and
mysterious intrusions of ocean. The visitor who, in 1911, picked
up those immature fish recorded that, while the boat in which
he was being rowed moved merrily at some distance from the cliffs,
and a piper seated in the bow played lustily, the piper stopped
playing whenever the boat, at the visitor's request, was brought
in toward the beetling face of Aoineag. Indeed, the piper was
actually overheard to mutter in the Gaelic to the Virgin Mary a
prayer for his deliverance. The natives of the Barra Isles, I should
remind you, are Catholics.

Mingulay, of course, is still remote, even although today so
many of the local fishermen possess their own motor-boats. These
they use chiefly in lobster-fishing, a pursuit now highly remunera-
tive owing to improved transport to London and other mainland
markets. Almost any day during the appropriate season, motor-
boats are operating profitably in the waters of this romantic group
of isles. The advent of this type of sea-transport has changed
things considerably in the Hebrides. In the days prior to the
Second World War, it was necessary to have one or two shepherds
resident on Mingulay for varying periods. Now, with the ubiquity
and manoeuvrability of the small, sea-going motor-vessel, one or
more shepherds, in reasonably good weather, can reach Mingulay
in a couple of hours, and can do with the sheep there, during a
long and active day ashore, all that is required. Gathering and
culling operations, and the bringing away from the island of the
lambs, take place late in August, or early in September. These

operations on Berneray are much easier than on Mingulay, where,
owing to such difficulties as are presented by the landing-places,
where such exist, every animal has to be manhandled individually.
Therefore, the day for attending to the sheep on Mingulay must
be chosen very carefully. The wind, particularly, has to be studied,
as regards its *direction* rather than as regards its velocity.

When Mingulay, along with Berneray and Pabbay, was offered
for sale in 1951, the particulars supplied to prospective purchasers included three possible landing-places on it, one at the
north-east of the island, one at the west side, and one at the south.
This, it was stated, rendered feasible a sheltered landing in any
wind. The drawback, however, was the absence of a pier, and
the none too serviceable condition of the boat-slip up which the
last inhabitants had been in the habit of hauling their craft out
of the sea by a winch now sorely in need of replacement.

This brings us to Mingulay in its peopled days. In 1861 its
population was 139, while Berneray's was 34. No figures appear
to be available for Pabbay at that time. By 1881, Mingulay's
population had risen to 150, all but 4 of whom were Gaelic-
speaking. By 1904 it had fallen to 135. That of Berneray was
then 17—just half what it had been in 1861. The steep decline
had now set in where islands such as these were concerned.
Pabbay had 11 inhabitants in 1904. The census of 1911 showed
that number to have fallen to 5. That year, such of the natives
of Mingulay as had not already betaken themselves to Vatersay
in 1907, when they squatted there, on the MacDonalds' farm,
joined their kinsfolk there.

Judging even by such scant accounts of Mingulay as exist,
its heyday (if such it might be termed) would appear to have
been somewhere about 1880. At that time, when, as we have seen,
its population was 150, it grazed a few hundred sheep, some
cattle, and a score of hardy shelties—of Highland ponies. 'The
ponies only have about a fortnight's work to do in each season,
carrying down the peat cut high up upon the hills,' wrote Harvie-
Brown. The inhabitants, in common with their neighbours on
Berneray, prosecuted lobster-fishing remuneratively, supplementing in this way a livelihood derived from crofting in the old,
traditional style. Their diet they certainly could have varied to
some extent, had they appreciated that many a thing they avoided,

LOCH LOMOND. Creinch (*left*) and Torrinch (*right*) from the high ground behind Ross Priory. In the background are the Luss Hills

LOCH LOMOND. Inch Cailleach from MacFarlane's Boatyard at Balmaha

MINGULAY. The 'Black Houses' comprising Mingulay village – the village as it was at the beginning of the present century

in the belief that it was poisonous, was edible. For instance, the immense quantities of mushrooms growing over toward the north-east of their island they regarded as something deadly, referring to them contemptuously in Gaelic as "cow-dung spots".'

Fuel in the form of peat was plentiful; and new hags were constantly being opened up as the older ones became exhausted. One may be excused for introducing here a not altogether inapposite ornithological note. The change from the old hags to the new immediately influenced the local distribution of the storm-petrel. This bird, which hitherto had bred in the dry cracks of the old peat-face, was often found there by the islanders when cutting their peats. Now they vanished from the scene. Mr. Finlayson, schoolmaster on Mingulay at the time of Harvie-Brown's and Buckley's visit in 1887, told these naturalists that for a great number of years he had not heard of a petrel's egg having been found where, formerly, they had been plentiful, 'though he does still occasionally see a bird in the twilight'.

That life on Mingulay then was idyllic, one can imagine from some of the photographs taken early in the present century by my very dear friend, the late Robert M. Adam, that supreme photographer of the Scottish scene, to whom I devoted a chapter in a recent book.[1] The loveliness of Mingulay Bay on a summer's day has to be seen—has to be *experienced*—to be believed. Many a traveller among the remoter isles of Britain has declared it to be finer than Village Bay, St. Kilda's. Of course, it *is* finer, as also somewhat less remote. The boulder-strewn storm-beach at Village Bay can be as repelling as the white sands of Mingulay Bay can be magnetic. What gives Village Bay its splendour, when viewed from the sea, is the lofty cirque forming its background.

However halcyon the summer on Mingulay, it compensated little for the hardship and privation the natives endured throughout the long and turbulent winter, when isolation was virtually as complete as on St. Kilda. The only thing of social value which the long, dark, wintry evenings encouraged was the *ceilidh*, with its traditional storytelling. On such occasions the island's folklore, its legends and traditions, were recounted in the glow of peatfire and cruisie, the latter being the old, open, iron oil-lamp used with

[1] *The Enchanted Isles: Hebridean Portraits & Memories*, published by Michael Joseph in 1967.

a rush wick. The older natives, renowned for their folk-memories, would recite the experiences of forefathers declared to have fought with The Bruce at Bannockburn and with the Jacobites at Culloden. Their participation in the former engagement seems improbable; but not so in the latter. The islanders were particularly fond of rehearsing the grim tale of an ancestor who, having survived the rout of Prince Charlie's forces at Drummossie, returned to Mingulay, believing himself to be safe from danger once more. But one day there appeared in the bay a vessel of war, the crew of which came ashore, seized the old campaigner, and hanged him on the ridge-pole of his cottage. Among those crofters who, early in the present century, quitted Mingulay to squat on Vatersay was an old man who claimed to have been the direct descendant of this unfortunate Jacobite. In quivering accents, and with tears flooding his eyes, he would recount this incident to any stranger landing on Mingulay, as though this ancestor of his had suffered ignominious death but a day or two previously. Many of the tales of Mingulay were concerned with tragedy; and I may, perhaps, be excused for including here one such example from my book, *Summer Days among the Western Isles*, published as long ago as 1929.

Once upon a time a certain chief of the MacNeils of Barra sent a steward named MacPhee to Mingulay for the purpose of collecting the rents due by his tenants there. At this period it was customary to have in the centre of every boat sailing any distance a peatfire in a three-legged pot, so that, in the event of accident or delay, food might be prepared at sea. On account of the difficulty usually experienced in effecting a landing on Mingulay, only MacPhee disembarked there. The crew remained aboard to keep the boat a little offshore until he returned. When MacPhee reached the dwellings of MacNeil's tenants, he was horror-stricken at finding all the inhabitants dead in their houses, as though some terrible catastrophe had overtaken them. He immediately fled back to the boat to summon assistance; but the crew, suspecting from what he had said that the people of the island must have perished of plague, refused to come sufficiently close to the shore again to enable him to re-enter the boat, for it was feared that he might have been carrying infection. And so, after having thrown out to him a smouldering peat, so that he might kindle a fire for himself, they set sail for Castlebay with the gruesome tidings that on Mingulay the steward had found not a living soul. It was explained that he had been left behind for fear of his smiting the natives of Barra with the plague that they surmised had wiped out Mingulay's population.

For a tale of Mingulay in times more recent, I revert once more to the Coddie, about whom I relate so much in *The Enchanted Isles*, and whose repertoire was by no means confined to the distant past. 'I can turn on the tap about the present as good as yourself, Alasdair Alpin!' he assured me more than once.

'Well, then,' I answered, 'why not reminisce a bit about your contemporaries in days gone by? What do you recall about the folk on Mingulay, who left there when you must have been in your early thirties?'

'Many a time yourself was on Vatersay,' the Coddie continued. 'And you saw there all the folk I knew years ago when I was a young man—many *cailleachs*[1] among them, and many *bodachs*[2] too, on Vatersay today. I heard you were in the MacLeans' house there, and saw the old spinning-wheel that spun the Mingulay wool in olden times, and Mrs. MacLean herself working it for you at the side of the fire for a photo myself saw. You would be sending the MacLeans a copy, I'm sure; and that would be the one myself saw. That's a whilie back. Och, yes, a good whilie back —before the *second* war. All the old people you and I knew then, *A bhalaich!*[3] are well in extinction now. They'll be lying in the *cladh*[4], for certain. Let me slap my box and see what comes out of it in the old memory line. I can tell you a good story about myself and John Finlayson, the Mingulay schoolmaster. John was a classmate of my own.'

The Coddie was never known to have embarked on the subject of John Finlayson without prefacing what he had to say by mentioning a memorable occasion on which the two of them were line-fishing off Eriskay, and by repeating an equally memorable fragment of their conversation in the boat.

' "Well, now," says myself to John, "it's a day of days this we're having together! A day of days, my boy! And all to our-

[1] old women.
[2] old men.
[3] My boy!
[4] graveyard.

selves, fishing here and there, and not a worry about the world."
Never a soul to put a hurry-burry on us. We would just be taking
our time. No one to order us to do this thing and that thing.

'Says John to myself, "*Se, gu dearbh!*[1] And there's myself on
Mingulay, spending days and days like this one, trying to insert
brains in small heads when the Almighty Himself didn't do it!"'

The Coddie, having divested himself of this picturesque
preliminary, then proceeded to his account of what befell John
when returning to Mingulay from Castlebay one Saturday evening
alone, in his own boat, and not a soul with him at the time.

'Tell me, Coddie! Has the day of the week anything to do with
what you're about to relate?' I enquired.

'Maybe, and maybe not. You will be better judging that for
yourself when you hear the story right. Well, John was getting
on fine with the homeward journey. He was as good a seaman as
myself. He got his Master's Ticket when he was a young fellow;
and no one in Mingulay—nor nobody in Barra itself—had a
better seamanship at his finger-ends. Often he sailed across to see
myself. He would be putting up a wee notice on the schoolhouse
when he would be going away like that. "School closed for the
weekend. Back maybe Monday or maybe Tuesday." He liked to
give himself plenty time for getting back from Castlebay, just in
case a nice entertainment turned up, and he couldn't get back to
Mingulay.

'Well, now, he was getting on when, all of a sudden, the breath
went out of the wind on him. There wasn't a drop in all the wide
space around him. So he had to wait. "Maybe I'll get a bit of a
blow before long," he says to himself. But never a blow did he get.

'The night was coming close to him, and the rocks was coming
close too. "*O vo vo!* boy! you'll have to get going with the oars!"
Well, he would be taking off some of his clothes with the heat
that was in him after rowing a mile or two. And what happened
to poor John but he fell in! There he was, all alone in the ocean!

'Well, he was trying this way and he was trying that way to
get back in the boat. He couldn't climb back at all. And himself a
good seaman. "What will the scholars be thinking when they hear
what the schoolmaster's been up to?" says he to himself, and
getting very tired with the weight of his clothes in the water.

[1] Yes, to be sure!

' "Nothing for it, boy, but have another try!" So he got hold
of the rudder—he couldn't climb up it. So he got the rudder-pins
out of the hooks, and, Man! he flung the rudder into the boat.
Now, the wily bird that he was put his big toe—I should have
said he had his boots off long before he was in this awkward position
I'm telling you about—he put his big toe on the rudder hook, the
one under the water, and landed himself fair and square back in the
boat like a wet salmon that jamp out of the sea with a great leap.

'He had an awful job getting the rudder right again. But what
would the good folk on Mingulay be saying about the schoolmaster coming back from Castlebay waterlogged like that, and
on a Saturday night? Nothing for it, boy, but strip and put his
clothes to dry on the thofts. So there he was, rowing home to
Mingulay like the Wild Man of Borneo in Nature's raiment. He
hadn't such a great hurry on him now, whatever; but he knew
they would be wondering and wondering about the schoolmaster.
No hurry, boy! The clothes must dry a bittie yet!

'So he pulled away bare to the skin at the oars till he saw a
wee bit of the sands on the beach in the corner of an eye. So he
hove-to, and made himself respectable before landing. Then he
rowed ashore in style, with every man, woman, and child of the
township down to welcome him home, waiting to give him a
hand to beach the boat.

' "Well, boys," says John to his friends, "you nearly lost your
schoolmaster tonight in the Minch there. I went over the side,
and couldn't get back in my clothes. So I took the rudder off its
sockets, and put my big toe down on the one at the bottom; and
then I got aboard that way. And here, with the help of Providence,
your schoolmaster is back among you all from the jaws of the
ocean, and thankful."

'There was a great rejoicing in Mingulay that night; and someone went away up to the schoolhouse—an advance-party like—to
put a peat or two on the embers. And John says to myself:
"Coddie, it's myself that mustn't tempt Providence more. I'll
have to be careful for the future when I go to Castlebay, and
myself coming home alone in the boat." '

*   *   *

The problem which faced the St. Kildans and brought about

the evacuation of Hirta in 1930 was very similar to that confronting the natives of Mingulay two or three decades earlier. It found
a solution when, in 1907, a number of crofters and cottars from
the island (on which at the time there were twenty crofts) squatted
on the farm of Vatersay, as already mentioned, and built thereon
homes in the nature of wooden hutments. Two or three years
later, the Congested Districts Board, forerunner of the Board of
Agriculture for Scotland, purchased Vatersay, and formed 58
holdings there. A third of these was allocated to people from
Mingulay, the last of whom left that island for Vatersay in 1911.
Mingulay at that time was owned by Lady Gordon Cathcart,
whom the natives looked upon as a wicked ogre, recalling in venge-
ful tones the cruel excesses alleged to have been perpetrated among
the Barra Isles by her father, Colonel Gordon of Cluny, at the
woeful time of the Clearances.

One of my first visits to Vatersay, early in the 1920s, took
me by invitation to the homes of some of those who had come
there from Mingulay. It was then that I heard at first hand about
life on Mingulay as it had been lived, with little perceptible
change, over a great number of years. The old folk on Vatersay,
still pining a little for their island, spoke of it with deep emotion.
Though now well and truly settled for more than a decade in
circumstances incomparably more favourable than those they
had left, their thoughts were never long absent from *Eilean mo
Chridhe*—from the Isle of their Hearts. They required no encouragement to describe its virtues, forgetting, for a while at least,
the privations they had endured there.

A visit I was able to make to Mingulay about this time was
of too short duration to allow of my exploring it as I would have
wished. In any case, the gales and blinding rains accompanying
me would have prevented my seeing much of the island, even if
I had been in a position to remain longer on that occasion. In
1937, however, Mingulay, Berneray, and Pabbay were purchased
by my friend, Peggy Greer, then farming, after a fashion, in
Essex, and now bent on farming Mingulay as it had never been
farmed before! A decade later, this somewhat misguided enterprise of hers was to be the means of my re-visiting Mingulay
under auspices that could have been singularly advantageous,
had Peggy been less vague—less Irish, as it were—about things.

In the summer of 1949, then, I joined her at Castlebay while she was contemplating a Sunday-afternoon visit to Mingulay aboard her own bulky motor-craft. She had in her employment at the time a certain Alick MacLean, a native of Castlebay. Alick was now her boatman and general factotum. In these capacities he supervised her sheep-stock on Mingulay from time to time. On reaching at Castlebay the pier whence we were to sail for Mingulay, I perceived at once that, from my own selfish point of view, the excursion on which we were about to embark was likely to be a disappointing one. Alick had invited several of his friends to accompany us; and those whom he hadn't invited had just invited themselves, in that nonchalant way in which the West Highlanders and Islanders are so adept. Peggy, who wasn't consulted, had no idea who they were, these people stepping aboard her boat at Castlebay that afternoon, as though she were obliged to provide free public transport for them.

We were well out to sea before Peggy, swithering hitherto as to whether she ought to say something, and, in so doing, risk offending Alick MacLean, upon whose quite generously remunerated services she had allowed herself to become all too dependent—Yes, we were well out in the Minch, halfway to Mingulay, before she mustered sufficient courage to enquire of this one who *he* was, and of that one who *she* was!

At this stage, *I* began to take a hand in things, making it clear to everybody aboard that, when we reached Mingulay, I did not want this trippery boat-load ranging over the island. In such time as might be available to me, I must cover what I could of it with my camera, and not be held up by people stravaiging into pictures which I wished to show as scenes forsaken by humanity some forty years earlier. I insisted that, apart from Peggy herself and Alick MacLean and a fellow invited with a view to his checking over the motor-plough landed on Mingulay some weeks previously, nobody should wander farther than a hundred yards from the boat until I had been at least two hours afoot on the island. Thereafter, by which time I expected to be climbing steeply to the edge of its western cliffs, they could do as they pleased.

The several hours I spent on Mingulay that distracting Sunday afternoon were as physically strenuous as any I have spent anywhere in my life. Determined to obtain what photographs I

could, even at the risk of returning to the boat at sundown in a
state of exhaustion, I dealt first with Mingulay's ruined township.
The sands of time, as I soon perceived, were slowly, but inexorably,
encroaching upon it. I realised that, unless a powerful, seaward
wind swept it unremittingly for some days, it would be completely
overwhelmed in a few years' time, and perhaps even difficult to
locate. The sands were steadily creating archaeology for succeeding
generations.

I then identified among the ruins the island boatbuilder's
house, and also the cradle where, in olden days, the keel of many
a small fishing-boat had been laid. Then to the old mill, driven
by a stream from the hills, its lade choked with irises, its
ponderous wheels of hewn stone buried, as were most of the
querns, the latter near thresholds silent and decayed. And then to
the island's primitive graveyard, readily recognisable by reason
of its little gravestones—stones of no memorial significance to
anybody except to those descendants on Vatersay. They, and
they alone, know whom these rude emblems were designed to
commemorate.

The only buildings on Mingulay which were still habitable, and
in any sense in a state of repair, were the schoolhouse and the
Priest's House. I render the latter with capitals, as the islanders
like to do. This is a two-storeyed place with a chapel on the first
floor, which explains why it is sometimes referred to among the
Catholics of the Barra Isles as the Chapel House. Well built of
squared granite and pitch-pine, it may be regarded as an early
example of modern planning. It has four rooms and a kitchen
downstairs. Above these, and reached by an external staircase,
is an apartment 45 feet by 25. This would have been the natives'
place of worship, had they not left before its completion. Among
Mingulay's last inhabitants were the workmen employed in
putting the finishing touches to this structure.

Having devoted what attention I could to the site of the
township and to the area once cultivated in its proximity, I left
Peggy discussing with Alick MacLean and his friend the efficiency
of the motor-plough she had imported with a view to reclaiming
at least part of the arable croft-lands that had lain fallow since
the last inhabitants had tilled them so many years earlier. Of
an area of roughly fifty acres formerly given over to potatoes,

only two or three had been ploughed up with this contrivance, and duly fenced in with potato-planting in prospect. How Peggy proposed disposing of the crop anticipated, I never knew. She fancied, I think, that there was a ready market for it in Castlebay, although everybody there already had his own potato patch! But Mingulay potatoes! Ah! who would eat Barra potatoes if he could get Mingulay ones! Well, the answer is NOBODY, unless you grew them, shipped them, cleaned them, and deposited them free of charge on the islanders' doorsteps in Barra!

I now made for the western cliffs, aware that their bird population would have been somewhat reduced by this time. The hatching season, for at least some species, was virtually over. But I had no doubt that young birds on the nest would still be fairly numerous. By the cliff-top, some hundreds of feet above the ocean, I lay a considerable time, watching a pair of greater black-backs systematically patrolling the cliffs' ledges below, in the hope of carrying off fat fledglings left temporarily unprotected. The black-backs' marauding occasioned a deafening uproar as parent birds, half their size and strength, returned to drive them off with flapping wings and pecking beaks. Judging by the hoarse croak a black-back emitted, it sounded as though, now and again, he had sustained a painful peck. The tumult thus created was aggravated considerably when the marauders winged past a colony of kittiwakes. How plaintive and piteous were the *kitti-waak-kitti-waak* cries of those members of the colony directly threatened by these black-backs! But order was soon restored after they had gone.

In lying flat on my stomach and drawing myself forward to the truly terrifying edge of the cliffs, I was able to locate young kittiwakes now delivering themselves of chirpings and cheepings that sounded as though they were an expression of thankfulness that the danger of their being devoured was past, if only very temporarily.

No less absorbing was an adjacent colony of herring gulls, most of its parent birds engaged in getting their young off the nests and on to the wing for the first time. They poked them. They prodded them. They shoved them forward to the edge of their nests, but seemed a little dilatory about tipping them over. Failure to achieve this final purpose in no way deterred them, however. They went on trying until, at last, a youngster was dislodged. This immediately sent a wave of encouragement through the entire gull colony. Parents, hitherto not too vigorous in their endeavours, now began ejecting their offspring with determination. One watched while a particular ledge was being cleared in this way of its fine, plump nestlings. Down below, on a sea gently crinkled, these young birds, so very recently cast upon the world to fend for themselves, floated in dense packs, feeling beneath them, for the first time in their precarious lives, the unceasing pulse of the ocean which was now to be their home until, in due season, their own nesting-time arrived.

Nobody is ever too fond of the gull, so common and undistinguished is he, so rapacious, so destructive of smaller and weaker birds. My own distaste for him arises from personal observation of the toll he takes of the shearwaters on the islands off the Pembrokeshire coast, particularly on Skokholm. More than once I have witnessed his slaughter of this wonder-bird. A few years ago I saw the same thing during some weeks' stay on King Island, situated in Bass Strait, roughly midway between the Australian mainland and Tasmania. There the greedy and ferocious gulls, for the sake of slaying, slew in great numbers the mutton-birds—the Australian short-tailed shearwaters, so called because their flesh is said to taste like mutton.[1]

Yet, on re-reading one of Dr. Mary Andrews's little books about Filey, as she knew it in her youth, I was to discover that the gull tribe is not entirely without its champions. Mary, who was doctor at the Derbyshire village of Shatton for many years, and whom I knew well, had been in the habit of watching the gulls with a pair of powerful binoculars. 'One evening,' she wrote, 'a solitary little chap spent much time quite alone by the tide edge. His two mates had been preening their feathers and then flown away. My bird made no attempt to follow. He could spread

---

[1] The shearwater's homing achievements are astonishing, prodigious, uncanny. This was demonstrated in 1958, when several of them were removed from their burrows on Skokholm, and flown to the United States. There, at Boston harbour, they were released. In forty seconds they had oriented themselves correctly with the sun, and set a 3,000-mile course across the North Atlantic. Within $12\frac{1}{2}$ days, every one of them reached its own particular burrow on Skokholm, flying at an average speed of 255 miles a day. In whom could such a bird fail to inspire wonder and reverence? Fancy our bedevilling with nationalism and politics, with all our dirty, little tricks, a world so wondrous!

his wings and take a short walk now and then, but never attempted
to get airborne. Was it a case of oil? If so Heaven help him!
Then a charming thing happened. A gull friend swooped towards
him, put down his 'undercarriage' and made a perfect landing
exactly alongside. Through the glasses I had a close-up of what
followed. They looked at each other, and I seemed to hear the
anxious enquiry—"What's wrong, old pal? Engine trouble—poor
take-off, not enough boost—or just wing trouble?"

'The newcomer looked full of sympathy with his little beady
eyes. They both preened feathers and sat down close together.
Then, presently, they walked away along the creamy sea fringe,
and passed off the screen as far as I was concerned. I cannot help
wondering just what happened to the little chap who was
apparently grounded, and if he escaped the big, black dog on his
lawless evening gallop over the sands. He has a mania for chasing
the seagull folk.'[1]

I must say that, ruthless as we know certain species of gull
to be, the Hebridean scene would be incomplete if he were
banished from it. What would Castlebay be like if he and his
ominous *cag-cag* were driven forever from its chimney-pots? Who
can visualise Stornoway's quaysides without gulls, or that part
of the town known as Newton, where, all day long, they pace up
and down the sea-wall, waiting for its housewives to tip their
garbage over it into the sea?

In May, 1959, the manner in which a dead gull was disposed
of at Sheshader, a crofting township no distance from Stornoway,
brought a native of the adjoining township of Shader before the
sheriff-court at Stornoway. The *faoileag,* as a little gull is called
in the Gaelic, was sitting innocently enough on a croft at Sheshader, puffing out his white chest and minding his own business,
as was reported in *The Scotsman.* Suddenly, an explosion rent
the air. A shotgun had blown him into eternity. His remains, we
were told, were carefully deposited by the wayside, immediately
in front of Annie MacLeod's croft, at Number 24, Sheshader.
Though the *faoileag's* death had occurred about the middle of
February, the incident did not receive statutory attention until
nearly three months later, when John MacMillan, a shopkeeper at

---

[1] *The Heart of Filey,* published by W. H. Lead, Ltd., Silver Street,
Leicester.

Shader, appeared before the sheriff, and pleaded Not Guilty to the charge that, in an open-air place to which the public has access without payment, contrary to the provisions of the Litter Act, which had become law the previous year, he had thrown down litter, 'namely, a dead seagull'.

According to the evidence of the police, MacMillan, having shot the seagull, deposited it near his van, and averred that he had arranged with Annie MacLeod's sister, Isobel, that she should remove it. But, since the corpse continued to lie just where he had thrown it down, Annie complained to the local policeman. 'Since MacMillan had put it there, I asked him to take it away,' she declared; 'and he told me he would do so the next time he was passing.' Her sister denied having given MacMillan any undertaking that *she* would remove the offending carcass. She had actually done so in the end simply because she disliked its being left rotting by the roadside. MacMillan was acquitted. The court upheld his solicitor's plea that he had merely set the gull down by the road prior to its disposal, and had not acted, therefore, in contravention of the Act!

But we must revert to Mingulay, our theme.

As time was getting on, I felt that by now Peggy Greer must have completed her survey, and come to sundry decisions regarding both prospective crops and the sheep-stock. This meant that she was ready to embark again. Far away, on the sands of Mingulay Bay, I could discern, from my solitary position at the cliff-top, a human figure or two. Reluctantly, I began to descend the valley toward the ruined village. I had not proceeded far when I heard that shrill whistle sheep-folk emit when directing their collies on some distant hillside. This was intended for me, of course. Realising that further concentration, either upon my notes or upon my camera, was now virtually impossible, I dropped speedily to the shore, consoling myself with the thought that I must have obtained a few photographs worthy of reproduction. How many miles on Mingulay I had travelled, I could not say; but I do know that, by the time I stepped aboard again, weariness of limb dissipated, in some measure at least, my disappointment at being unable to remain on the island another hour or two. This would have been practicable without encroaching seriously on the length of daylight necessary for our safe return to Castlebay. I took one final photograph of the fallen and deserted township, the sun by now having declined sufficiently to cast long shadows across its hollow ruins.

If, perhaps, I felt a little frustrated by circumstances over which I had so little control, I was promptly restored to equanimity when Alick MacLean, on Peggy's instruction, undertook to take me back to Mingulay a few days later when, in any case, he would have reason to return there. The conditions promised were to be infinitely more conducive from *my* point of view than those obtaining that day. To begin with, we would sail from Castlebay *alone*. Furthermore, Alick would be so occupied with fencing operations on Mingulay that I could have at my disposal as many hours of daylight as I desired. We would bring food with us, and also bedding, so that, in the event of either of us not having completed to his satisfaction the particular task on which he was engaged, we could spend the night in reasonable comfort in the Priest's House, the key of which was in Peggy's possession. The only thing which might interfere with this consoling arrangement was an unpropitious change of weather.

But the weather held. Indeed, from the camera's viewpoint, it improved. With this prospect of re-visiting Mingulay, I altered my plans by telegraph and telephone, deciding to delay my departure from Castlebay for a few days. Two evenings later, Alick MacLean informed me that all was ready for our return the following morning. He already had stowed aboard Peggy's motorboat the fencing material required. The wireless weather forecast for the Hebrides that evening was highly auspicious. 'You'll get all the photos. you want on Mingulay tomorrow,' he assured me. 'Just our two selves going. No distractions this time. I'll be getting on with the fence, while you'll be away over the hills with the camera.'

The morrow broke clear. At the hour agreed, I presented myself at Alick's doorstep. But there wasn't a sign of Alick anywhere.

'Where's Alick?' I asked his mother. 'He was to take me to Mingulay today.'

'Och, well, I don't know where Alick is at aall, at aall. A wee stirkie got away on them. It's away up the hills they'll be, catching the stirkie. It got away on them. Och, they'll be back, though. They'll be back aall right. They'll get the stirkie, I'm sure.'

Mrs. MacLean could give me no idea how long her son would be gone on this wild-stirkie chase. Nor could she say that she knew he intended taking me to Mingulay that day. I decided to return to the MacLeans' cottage at noon.

'They're away after the stirkie that got away on them!' was the twelve-o'clock news. 'Och, they'll get the stirkie aall right. Alick'll be back sometime—when they catch the wee stirkie.'

Enquiry by the pier at Castlebay elicited that a panic-stricken beast, while being loaded on to a boat, had leapt overboard, had swum ashore, and had bolted to freedom among the hills of Barra. True to custom in such circumstances, Alick had joined its pursuers. Not a word to me that he was off on a day's chase, though I was living within a few minutes' walk of his cottage. Such civilities, even in the Hebrides, are outmoded. So I decided to waste no further time among the unreliable and procrastinating Celts. Resolved to make the most of a promised spell of fine weather, I flew from Barra to Glasgow that afternoon, and thence to Shetland. Thus ended my attempt to see more of Mingulay.

That stirk at large on Barra reminds me of the pranks of another of his tribe on Mingulay. When on one occasion Seton Gordon, the naturalist, landed there with his provisions, intent on remaining some days, he was welcomed by a stirk that had been brought up as a pet on one of the other islands, and had been shipped to Mingulay, with others, for grazing. Glad of the opportunity of renewing acquaintance with humanity, such as it had been used to in its juvenile days, the animal charged Seton in an irresponsible way, knocked him over, and in the process broke all the eggs he had brought with him as part of his provender.

*   *   *

Isle of my Heart, so dear to the few still alive who, sixty years ago, left their native hearths for more congenial surroundings, remained in Peggy Greer's possession, along with Berneray and Pabbay, until 1951, when she decided to dispose of them. That year, in *The Times*, an attractive advertisement, 'Islands for Sale', caught my eye. The islands turned out to be Peggy's! The advertisement ran as follows:

'Islands for Sale, 2,500 acres, with boat, two houses, and sheep at valuation.'

The prospectus, promptly sent to enquirers on application, requested them to communicate with Messrs. Skene, Edwards, & Garson, 5, Albyn Place, Edinburgh. It stressed the islands' advantages. MacBrayne's mailboat connections with Barra were suitably underlined, as was also British European Airways' daily flight from Renfrew airport to the sands of Barra's Great Cockle Shore. The islands' potentiality as a profitable sheep and cattle run was likewise emphasised:

The Gulf Stream flows on both sides of them, so the climate is mild. Snow never settles; and stock can live out all the year, with no artificial feeding. So no hay nor silage-making is required, and no bought foodstuffs. The Islands have plenty of shelter. There is little bracken and plenty of wild white clover. At present they are understocked with approximately 400 ewes and gimmers, plus 5 rams, all Blackfaces of a well-known pure strain. Pabbay is also highly suited to cattle, and would take 80 head, plus 150 sheep, without strain on the grazing. A full sheep stock of the 3 islands would be over 1,000 —800 on Mingulay, 200 on Berneray, 150, plus cattle, on Pabbay. The average nett income of 20 sheep in the Outer Hebrides is £50 a year.... There are no rates, the land-tax on all 3 islands being £26 2s. 6d. a year, & the Minister's Stipend £4 7s. 10d. a year.... There is shooting in the autumn —woodcock, snipe, and wild geese—but no fishing other than in the sea, where lobsters are plentiful. Basking sharks, huge but harmless, are common at some seasons; while seals breed in places on the Islands. There are no trees, though fruit bushes were formerly grown by the islanders; and there are many wild flowers, especially primroses. I will be glad to answer further enquiries on behalf of the Owner, Mrs. P. B. Greer, while *bona fide* buyers are referred to her solicitors, Messrs. Skene, Edwards, & Garson. I ask all whose interest in Islands is purely theoretical not to waste the time of solicitors or myself.'

Withal, an immensely desirable property—on paper, at any rate! Although I knew that Peggy, with her own literary qualifications, must have had a hand in framing this glowing prospectus, I was greatly interested in 'myself' who authorised it. I discovered that he was, and still is, the versatile Lawrence D. Hills, who lives near Braintree, and at the present time is, *inter alia,* honorary secretary of the Henry Doubleday Research Association. Lawrence's name became familiar to many in connection with his regular contributions on gardening topics to the columns of *The Observer* over a number of years. He himself has described to me all that ensued from his insertion in *The Times* of that advertisement. More than a hundred replies were received. Roughly a dozen titled people, including a couple of peeresses, applied. An Oxford don thought *he* would like the islands. So did a fleet of Commanders, R.N. (retired). So also did a mess of Colonels and Majors. Without exception, these were applicants who 'want to get away from it all, and live on a desert island'. They were precisely the theoretical applicants Lawrence had envisaged, and of whom he had hoped to be free. But a number of Scottish farmers applied too. So also did an organisation describing itself as Cumberland and Westmorland hill-sheep breeders.

Peggy's reasons for wishing to retire from her islands was sound enough. She had bought them when in her vigorous fifties. The time had now arrived, she felt, when she ought to part with them, before their sheep-stock and pastures deteriorated on account of her inability to devote to them that personal supervision they demanded. Moreover, she had at last found for herself what, many years before, I had warned her about, namely, that, living eleven months of the year in Essex, she was entirely at the mercy of the whimsical West Highlanders where the management of her islands was concerned. It was all very fine her having been adjudged the best lady-ploughhand in all England in the over-seventy class! That didn't qualify her to sheep-farm these islands by proxy, and at a distance from them of between six and seven hundred awkward miles!

Peggy's experiences as an island-*seller,* rather than as an island-*buyer,* or, rather, Lawrence D. Hills's experiences on her behalf, were indeed illuminating. From the moment he drafted that advertisement, he realised that any estate agent can let an unfurnished flat anywhere in the world, and that there exists any number of firms capable of selling anything from a bungalow the size of a dog-kennel to a mansion. But islands! Ah! islands! This was something very different. The selling of islands, as he quickly appreciated, was an art of its own.

MINGULAY. Graveyard of a vanished people. Note the rude tombstone and crosses

MINGULAY. The sands of Mingulay Bay encroach upon the ruined township

MINGULAY Bay at 10 p.m. on Midsummer's day

MINGULAY. Peggy Greer inspecting the old croft-lands where she planted potatoes in 1949. In the background stands the Priest's House

To every applicant he sent, in addition to the formal 'order to
view', a fuller and more truthful account of the properties involved
than house-agents are even *expected* to supply. This he did because
he knew that anybody seriously considering the purchase of these
islands would have to spend quite a bit of money in getting there
to inspect them. For one reason or another, only the merest
fraction of the applicants actually got as far as the islands. Of
three yachtsmen who tried to reach them, only one came within
landing-distance when, during a brief spell of calm weather, he
hove-to off Berneray. But the Barra Head lightkeepers got him
on his radio in time to save him from attempting what *they* would
not have attempted even aboard the powerful lighthouse tender.

Most of the would-be purchasers who got as far as Barra soon
got bogged down at Castlebay, in the bar at its hotel. When the
seas abated, a few of them did manage to hire a local boat to take
them to Mingulay. Among these were one or two who afterwards
put in offers of less than five shillings an acre. None of them, in
Peggy's view, was the least competent to farm the islands; and,
since she was anxious that they should go to a good home, as it
were, each and every offer was turned down. One would require
to know something more than just the ins and outs of sheep-
farming to deal successfully with these islands. Any new owner
would have to be able to handle a boat among islands almost
entirely shoreless, situated in perilous seas, if not also in faeryland
forlorn.

As Peggy Greer did not succeed in disposing of her islands in
1951, the same agents offered them for sale again in 1954. The
following year they were purchased by the partnership of five
residents on Barra known as the Barrahead Sheep Stock Company.
Mingulay remains uninhabited as before, although the new owners,
landing there from time to time to see to their sheep-stock, sometimes tarry a night or two in the Priest's House.

Mingulay! Try to pronounce the name as indicated near the
beginning of this chapter! These three syllables, euphoniously
rendered, unlock in the memory of those who have known this Isle
a treasure-house overflowing with something inestimably precious.

Index

MIDLOTHIAN COUNTY LIBRARY